化学工业出版社"十四五"普通高等教育规划教材

无机与分析化学实验
（双语版）

张丽影　王　茹　王怡婷　主编

Inorganic and Analytical Chemistry Experiments

化学工业出版社

·北京·

内容简介

《无机与分析化学实验》以无机与分析化学实验的基本原理为主线，注重基本操作和基本技能的训练。全书共分 5 个部分 23 个实验，内容包括：实验基本知识、基本操作、基础实验、综合拓展实验和附录。选编的 15 个基础实验主要涉及无机化合物的制备与提纯、化学平衡常数的测定以及利用化学滴定分析对物质进行含量测定。实验讲义以中英文对照方式编写，旨在扩大学生专业英语词汇量，便于双语教学。综合拓展实验融入无机化学学科发展前沿以及与生物、食品、材料、环境等专业相融合的内容，便于相关专业选做。

本书可作为高等院校化学、应化、化工、生物、制药、材料、食品、环境等专业的实验教材，也可供相关专业的研究及技术人员参考。

图书在版编目（CIP）数据

无机与分析化学实验：汉、英 / 张丽影，王茹，王怡婷主编. —北京：化学工业出版社，2024.8
ISBN 978-7-122-45718-9

Ⅰ.①无… Ⅱ.①张… ②王… ③王… Ⅲ.①无机化学-化学实验-汉、英 ②分析化学-化学实验-汉、英 Ⅳ.①O61-33②O652.1

中国国家版本馆 CIP 数据核字（2024）第 103371 号

责任编辑：满悦芝　石　磊　　　　文字编辑：曹　敏
责任校对：李露洁　　　　　　　　　装帧设计：张　辉

出版发行：化学工业出版社
　　　　（北京市东城区青年湖南街 13 号　邮政编码 100011）
印　　刷：三河市航远印刷有限公司
装　　订：三河市宇新装订厂
787mm×1092mm　1/16　印张 7½　字数 182 千字
2024 年 9 月北京第 1 版第 1 次印刷

购书咨询：010-64518888　　　　　售后服务：010-64518899
网　　址：http://www.cip.com.cn
凡购买本书，如有缺损质量问题，本社销售中心负责调换。

定　　价：28.00 元　　　　　　　　版权所有　违者必究

前言

无机与分析化学实验作为高等学校化学、化工及其相关专业的第一门基础化学实验课程，在加深学生对无机与分析化学理论的理解、培养学生实事求是的科学态度和良好的实验素养方面发挥着重要作用。

结合《普通高等学校本科专业类教学质量国家标准》《化学类专业化学实验教学建议内容》等文件，本书以无机与分析化学实验的基本原理为主线，注重基本操作和基本技能的训练；以"必需，够用"为原则，力求简洁，讲清概念，强调实用。全书分5部分共23个实验，内容包括：实验基本知识、基本操作、基础实验、综合拓展实验和附录。选编的15个基础实验主要涉及无机化合物的制备与提纯、化学平衡常数的测定以及利用化学滴定分析对物质进行含量测定。基础实验以双语对照方式编写，旨在提高学生专业英语水平，打开学生国际化视野。综合拓展实验部分，依托教师科研方向，以科教融合方式融入无机化学学科发展前沿，提高学生创新和应用能力；以学科融合方式融入生物、食品、材料、环境等专业内容，拓宽学生知识储备，同时便于相关专业进行选做。

本教材由张丽影、王茹和王怡婷主编。第一部分介绍了无机与分析化学实验的目的和意义、学习方法、实验室安全规则及数据处理方法，由张丽影、王茹、那立艳编写；第二部分包括无机与分析化学实验所需的基本器具和常规操作等，由王茹编写；第三部分15个基础实验，由张丽影、王怡婷、海华编写；第四部分综合拓展实验，由华瑞年、张丽影、王茹和王怡婷编写。附录由王茹编写。

全书由华瑞年教授主审。

由于编者水平有限，书中疏漏之处在所难免，热忱欢迎各位读者批评指正，以便及时更正。

编者

2024年7月

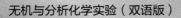

目录

第一章 实验基本知识 — 1

1.1 实验目的意义 — 1
1.2 实验学习方法 — 2
1.3 实验安全规则 — 4
1.4 实验数据处理 — 7

第二章 实验基本操作 — 14

2.1 基本器具 — 14
2.2 仪器的洗涤与干燥 — 19
2.3 容量仪器的使用及溶液配制 — 21
2.4 试剂及其取用 — 23
2.5 加热方法 — 25
2.6 蒸发、浓缩与结晶 — 26
2.7 固液分离 — 27
2.8 试纸的使用 — 29

第三章 基础实验 — 31

实验一 氯化钠的提纯 — 31
Experiment 1 Purification of Sodium Chloride — 33
实验二 硝酸钾的制备和提纯 — 34
Experiment 2 Preparation and Purification of Potassium Nitrate — 36

实验三　醋酸标准电离常数和电离度的测定 ·· 38
Experiment 3　Determination of Ionization Constant of Acetic Acid and Its
　　　　　　　Ionization Degree ·· 39
实验四　磺基水杨酸合铁（Ⅲ）配合物组成及稳定常数的测定 ····················· 41
Experiment 4　Determination of Composition and Stability Constant of
　　　　　　　Sulfosalicylic Acid Iron（Ⅲ）Complex ··· 43
实验五　$BaSO_4$溶度积常数的测定（电导率法） ··· 46
Experiment 5　Determination of the Solubility Product Constant of $BaSO_4$
　　　　　　　（Electrical Conductivity Method） ··· 48
实验六　碱金属和碱土金属元素鉴定 ·· 50
Experiment 6　Identification of Alkali Metal and Alkaline Earth
　　　　　　　Metal Elements ·· 52
实验七　硼族元素、碳族元素和氮族元素鉴定 ·· 54
Experiment 7　Identification of Boron，Carbon，and Nitrogen Group
　　　　　　　Elements ··· 57
实验八　氧族元素和卤族元素鉴定 ·· 60
Experiment 8　Identification of Oxygen and Halogen Elements ····················· 62
实验九　铬、锰、铁、钴、铜、银、锌、汞元素鉴定 ······································ 65
Experiment 9　Identification of Chromium，Manganese，Iron，Cobalt，
　　　　　　　Copper，Silver，Zinc and Mercury ··· 68
实验十　盐酸标准溶液的配制与标定 ·· 71
Experiment 10　Preparation and Standardization of Hydrochloride Acid
　　　　　　　　Standard Solution ··· 72
实验十一　混合碱中各组分含量的测定（双指示剂法） ·································· 74
Experiment 11　Determination of the Composition of Mixed Base
　　　　　　　　（Double-tracer Technique） ·· 76
实验十二　EDTA标准溶液的配制与标定 ··· 79
Experiment 12　Preparation and Standardization of EDTA Standard Solution ······· 81
实验十三　水中钙、镁含量的测定（配位滴定法） ·· 83
Experiment 13　Analysis of the Concentrations of Calcium and Magnesium
　　　　　　　　Ions in a Water Sample（Complexometric Titration Method）··· 84
实验十四　葡萄糖含量的测定（间接碘量法） ·· 86
Experiment 14　Determination of Glucose（Indirect Iodometry） ················ 88
实验十五　化学需氧量（COD）的测定（高锰酸钾法） ·································· 90
Experiment 15　Determination of Chemical Oxygen Demand（COD）
　　　　　　　　（$KMnO_4$ Titrimetry） ·· 91

第四章　综合拓展实验　94

实验十六　纳米硒的制备与表征　94
实验十七　上转换纳米粒子的制备与表征　95
实验十八　金属酞菁的制备与表征　96
实验十九　葡萄糖酸锌的制备与含量测定　98
实验二十　石墨相氮化碳量子点的制备与水质硫化物的测定　100
实验二十一　土壤中有效磷的测定　101
实验二十二　食用油中不饱和脂肪酸的测定　103
实验二十三　白酒中甲醇含量的测定　104

附　录　107

一、定性分析试液的配制方法　107
二、特殊试剂的配制方法　108
三、常用标准溶液的配制与标定　109
四、常见离子鉴定方法　110

参考文献　114

第一章
实验基本知识

化学是一门以实验作为基础的自然科学,任何化学规律的发现和化学理论的建立,都必须以严格的实验为基础,并受实验的检验。化学实验教学对培养学生的创新能力和优良的科学素养起着十分重要的作用。通过化学实验教学,可以强化学生的基本知识和基本理论,训练学生的基本操作和基本技能,培养学生分析问题和解决问题的能力,使学生养成良好的实验素养和严谨的科学态度。

1.1 实验目的意义

无机与分析化学实验是化学实验科学的重要分支,是基础化学实验平台的重要组成部分,也是高等工科院校化学、化工、生物、制药、环境、食品、能源、材料等专业的基础必修课程。它突破了原无机化学和分析化学实验分科设课的界限,使之融为一体。开设这门课程的主要目的如下。

(1)实验:通过实验获得感性知识,经分析、归纳、总结,从感性认识上升到理性认识,使课堂教授的重要理论和概念得到验证、巩固、充实和提高,并适当地扩大知识面。

(2)训练:正确掌握实验操作的各种基本技能和重要常用仪器的使用方法,培养独立工作能力和独立思考能力,为后续各类实验以及今后从事相关工作打下良好基础。

(3)观察、分析和处理:培养细致观察、及时记录实验现象以及归纳总结、正确处理数据、用文字表达结果的能力;培养分析实验结果的能力和一定的组织实验、科学研究和创新开发的能力。

(4)培养科学实验素养:通过实验逐步树立"实践第一"的观点,培养学生养成实事求是的科学态度和科学的逻辑思维方法,培养敬业、一丝不苟和团队协作的工作精神,养成良好的实验室工作习惯。

1.2 实验学习方法

做好无机与分析化学实验，达到上述实验目的，除应有正确的学习态度外，还需要正确的学习方法。简单总结如下。

1.2.1 预习

为使实验获得良好效果，实验前必须认真预习。

（1）认真阅读实验教材和理论课教材的相关内容并注意阅读其他相关参考资料。

（2）明确实验目的与要求，理解实验原理，了解操作步骤和注意事项，设计好数据记录格式。

（3）在认真预习的基础上，写出实验预习报告，其内容包括实验原理、实验步骤、操作要点和记录数据的表格。实验前预习报告要经指导教师检查后方可开始实验。

1.2.2 实验

（1）进入实验室后不急于动手实验，根据分组情况到指定位置实验台，仔细检查实验所用仪器设备，如有损坏应及时报告指导教师给予更换，未提前声明，责任由本次实验操作者承担。对于首次使用的仪器设备，经指导教师允许后方可使用。

（2）实验原则上应根据实验教材上所提示的方法、步骤和试剂进行操作，设计性实验或对一般实验提出的新实验方案，应与指导教师讨论、修改和定稿后方可进行实验。

（3）实验时，保持安静，遵守纪律，听从安排，认真仔细地完成每一步操作；实验过程中，应严格按规定进行操作，仔细观察并如实详细记录实验现象和数据。

（4）如发现实验现象和理论不符或实验结果误差过大，应认真分析并仔细查找原因，有疑问时力争自己解决问题，也可相互轻声讨论或询问指导教师或重做，直至获得满意结果；应随时记录实验数据，数据记录要实事求是，详细准确，不允许使用铅笔，不得随意涂改。

（5）实验结束后，在实验室使用记录本和仪器药品使用记录本上签字，将实验中使用的玻璃仪器清洗干净，整理药品，实验台上所有仪器恢复原状，排列整齐，经指导教师检查合格并在实验记录上签字后，方可离开实验室。

1.2.3 实验报告

撰写实验报告是实验的重要组成部分，同时也是评定该次实验成绩的重要依据。实验结束后，根据原始记录分析实验现象，整理实验数据。写好实验报告后，按照与指导教师约定时间及时上交实验报告。报告要求简明扼要，详略得当，叙述清楚，字迹清晰，整洁美观。

下面列出不同类型的实验报告形式，以供参考。

【例 1-1】 无机物制备实验

实验名称：××××的制备

姓名：_____　　同组人：_____　　日期：_____

一、实验目的（略）

二、实验原理（略）

三、实验步骤

（尽量采用表格、图表、符号等形式简明扼要地表示。）

四、实验数据处理（仔细观察、全面正确表达）

1. 产品外观
2. 理论产量和实际产量
3. 产率

五、问题与讨论

针对本次实验产生的实验现象，遇到的疑难问题或数据处理时出现的结果展开讨论，分析实验误差的原因，总结实验收获，也可对实验方法、教学方法、实验内容等提出自己的意见或建议。

【例 1-2】 测定实验

实验名称：××××的测定

姓名：_____　　同组人：_____　　日期：_____

一、实验目的（略）

二、实验原理（略）

三、实验步骤

（尽量采用表格、图表、符号等形式清晰明了地表示。）

四、实验数据记录与处理

若有数据计算，务必清楚表达所依据的公式和主要数据。以下以醋酸标准电离常数的测定为例说明。

1. 数据记录（完整、实事求是）

表×-× 醋酸标准电离常数和电离度的测定实验数据 ［室温(℃)＝　　］

序号	$c/(mol/L)$	pH	$[H^+]/(mol/L)$	$[Ac^-]/(mol/L)$	$K_a^\ominus$	α
1						
2						
3						
4						
5						

2. 数据处理（计算过程均需在实验报告上呈现）

在一定温度下，用酸度计测定已知浓度 HAc 的 pH 值，根据 $pH=-\lg[H^+]$，换算出 $[H^+]$，代入式（1-1）和式（1-2），求该温度下 HAc 的标准电离常数 $K_a^{\ominus}$ 和电离度 α。

$$K_a^{\ominus}=\frac{[H^+]^2}{cc^{\ominus}} \tag{1-1}$$

$$\alpha=[H^+]/c \tag{1-2}$$

五、问题与讨论

针对本次实验数据处理时出现的结果展开讨论，分析实验误差的原因，总结实验收获，也可对实验方法、教学方法、实验内容等提出自己的意见或建议。

【例 1-3】 性质实验

性质验证类实验以试管实验为主，小实验较多，内容广而杂。重点在于观察、记录和解释实验现象并加以归纳得出结论，因此实验报告的格式可做适当修改，采用表格式，尽量使用符号和化学方程式来说明反应机理。

实验名称：××××族元素性质

姓名：_____ 同组人：_____ 日期：_____

表格式：

序号	实验项目	实验步骤	实验现象	现象解释
1	实验项目名称	步骤简明扼要，尽量使用符号表示	是否有气体、沉淀生成，溶液颜色是否发生变化等	以化学反应方程的形式加以说明，某些情况可有多个反应方程式

1.3 实验安全规则

在化学实验中会接触许多有一定危险毒害性的化学试剂和易于损坏的仪器设备，如忽视安全问题、麻痹大意，则可能发生各种事故。因此对于初次进行化学实验的学生，必须进行安全教育学习，而且每次实验前都要仔细阅读实验室中的安全注意事项。在实验过程中，要遵守以下安全守则。

1.3.1 实验室使用规则

无机与分析化学实验是学生进入大学后学习的第一门基础实验课程。本实验的教学对于培养优良的实验素养起到至关重要的作用，因此必须严格要求自己，遵守实验室规则，养成良好的实验习惯。

① 严格遵守实验时间，不迟到，不早退，保持室内安静，未经指导教师同意擅自缺席者，取消本次实验成绩。

② 实验室工作必须保持严肃、严密、严格、严谨。室内保持整洁有序，不准喧哗、打闹，严禁吸烟、吃东西、喝饮料，不许随地吐痰、扔废物。

③ 进入实验室后，严格按照教师指定的实验台及分组进行实验，严禁互相交换实验台，如发生实验仪器、器材损坏或其他违纪行为，后果仍由该位置原指定的同学负责。

④ 实验过程中，节约使用药品、水电。废纸、火柴梗等倒入垃圾桶，废液分类别倒入废液桶，严禁倒入水槽，防止水池堵塞和腐蚀。有毒液体集中处理。

⑤ 实验结束后，将仪器清洗干净，放回规定位置，擦净实验台，仔细检查本实验台仪器设备是否恢复原状，并在实验室使用记录本及仪器药品使用记录本上做好记录，待指导教师检查合格并签字后，方可离开。

⑥ 实验全部结束后，由指导教师指定的同学担任值日生，负责实验室的卫生清扫及药品整理工作，最后检查水龙头是否关紧、电源闸刀是否断开。值日生整理完毕后，在值日生表上签字，经指导教师同意后，离开实验室。

1.3.2 化学试剂使用安全守则

化学试剂中的部分试剂具有易燃、易爆、腐蚀性或毒性等特性，化学试剂除使用时注意安全和按操作规程操作外，保管时也要注意安全，要防火、防水、防挥发、防曝光和防变质。化学试剂的保存，应根据试剂的毒性、易燃性、腐蚀性和潮解性等各不相同的特点，采用不同的保管方法。

① 一般单质和无机盐类的固体：应放在试剂柜内，无机试剂要与有机试剂分开存放，危险性试剂应严格管理，必须分类隔开放置，不能混放在一起。

② 易燃液体：主要是极易挥发成气体的有机溶剂、遇明火即燃烧。实验中常用的有苯、乙醇、乙醚和丙酮等，应单独存放于阴凉通风处，特别要注意远离火源。

③ 易燃固体：无机物中如硫黄、红磷、镁粉和铝粉等，着火点都很低，也应单独存放于通风干燥处。白磷在空气中可自燃，应保存在水里，并放于避光阴凉处。

④ 遇水燃烧的物品：金属锂、钠、钾、电石和锌粉等，可与水剧烈反应放出可燃性气体。锂要用石蜡密封，钠和钾应保存在煤油中，电石和锌粉等应放在干燥处。

⑤ 强氧化剂：氯酸钾、硝酸盐、过氧化物、高锰酸盐和重铬酸盐等都具有强氧化性，当受热、撞击或混入还原性物质时，可能引起爆炸。保存这类物质，一定不能与还原性物质或可燃物放在一起，应存放在阴凉通风处。

⑥ 见光分解的试剂（如硝酸银、高锰酸钾等），与空气接触易氧化的试剂（如氯化亚锡、硫酸亚铁等），都应存于棕色瓶中，并放在阴凉避光处。

⑦ 容易侵蚀玻璃的试剂：如氢氟酸、含氟盐、氢氧化钠等应保存在塑料瓶内。

⑧ 剧毒试剂：如氰化钾、三氧化二砷（砒霜）、氯化汞（升汞）等，应由专人妥善保管，取用时应严格做好记录，以免发生事故。

1.3.3 实验室安全防火守则

① 以防为主，杜绝火灾隐患。了解各类有关易燃易爆物品知识及消防知识，遵守各种防火规则。使用易挥发易燃液体试剂（如乙醚、丙酮、石油醚等）时，要保持室内通风良好。绝不可在明火附近倾倒、转移这类试剂。

② 进行加热、灼烧、蒸馏等操作时，必须严格遵守操作规程。易燃液体废液，要用专用容器收集后统一处理，绝不可直接倒入下水道，以免引发爆炸事故。

③ 加热易燃溶剂，必须用水浴或封闭电炉，严禁用灯焰或电炉直接加热。点燃煤气灯

时，应先关闭风门，后点火，再调节风量；停用时要先关闭风门，再关煤气，要防止煤气灯内燃。使用酒精灯时，灯内燃料最多不得超过灯体容积的 2/3。不足 1/4 时应先灭灯后再添酒精。点火时要用火柴或打火机点，绝不可用另一燃着的酒精灯去点。灭灯时要用灯帽盖灭，绝不可用嘴去吹，以免引起灯里酒精内燃。电炉不可直接放在木制实验台上长时间使用，加热设备周围严禁放置可燃、易燃物及挥发性易燃液体。

④ 加热试样或实验过程中小范围起火时，应立即用湿石棉布或湿抹布扑灭明火，并拔去电源插头，关闭总电闸。易燃液体（多为有机物）着火时，切不可用水去浇。范围较大的火情，应立即用消防沙、泡沫灭火器或干粉灭火器来扑灭。精密仪器起火，应用四氯化碳灭火器。实验室起火，不宜用水扑救。

⑤ 在实验室内、过道等处，需经常备有适宜的灭火材料，如消防沙、石棉布、毯子及各类灭火器等。消防沙要保持干燥。电线及电器设备起火时，必须先切断总电源开关，再用四氯化碳灭火器熄灭，并及时通知供电部门。不许用水或泡沫灭火器来扑灭燃烧的电线电器。人员衣服着火时，立即用毯子之类物品蒙盖在着火者身上灭火，必要时也可用水扑灭。但不宜慌张跑动，避免使气流流向燃烧的衣服，再使火焰增大。

1.3.4 实验室安全用电守则

① 用电线路和配置应由变电所维修部门安装检查，不得私自随意拉接。
② 专线专用，杜绝超负荷用电。
③ 使用烘箱、电炉等高热电器要有专人看守。恒温箱需经长时间试用检查，确定确实恒温后方可过夜使用。
④ 不用电器时必须拉闸断电或拔下插头。
⑤ 保险丝烧坏要查明原因，更换保险丝要符合规格，或通知变电所更换。
⑥ 经常检查电路、插头、插座，发现破损立即维修或更换。

1.3.5 实验室意外事故处理

实验过程中应十分注意安全，如发生意外事故可采取下列相应措施。
① 烫伤：可用高锰酸钾或苦味酸溶液擦洗灼烧处，再涂上凡士林或烫伤药膏。
② 受强酸腐蚀：立即用大量水冲洗，然后用碳酸氢钠溶液清洗，再用水冲洗，擦上凡士林。
③ 受强碱腐蚀：立即用大量水冲洗，然后用柠檬酸或硼酸饱和溶液清洗，再擦上凡士林。
④ 割伤：立即用药棉揩擦伤口，擦上紫药水用纱布包扎。
⑤ 毒气侵入：吸入有毒气体（如 CO、Cl_2、H_2S 等）而感到不舒服时，应及时到窗口或实验室外呼吸新鲜空气。

1.3.6 其他实验操作注意事项

① 洗液、强酸、强碱等具有强烈的腐蚀性，使用时应特别注意，不要溅在皮肤、衣服或鞋袜上。
② 有刺激性或有毒气体的实验，应在通风橱内进行，嗅闻气体时，应用手轻拂气体，把少量气体扇向自己再闻，不能将鼻孔直接对着瓶口。
③ 加热试管时，不要将试管口对着自己或他人，也不要俯视正在加热的液体，以免液

体溅出使自己受到伤害。

④ 有毒试剂（如氰化物、汞盐、铅盐、钡盐、重铬酸盐等）要严防进入口内或接触伤口，也不能随便倒入水槽，应倒入回收瓶回收处理。

⑤ 稀释浓硫酸时，应将浓硫酸慢慢注入水中，并不断搅动，切勿将水倒入浓硫酸中，以免迸溅，造成灼伤。

⑥ 禁止随意混合各种试剂药品，以免发生意外事故。

⑦ 实验完毕后，将实验台面整理干净，洗净双手，关闭水、电、气等阀门后方可离开实验室。

1.4 实验数据处理

在计量或测定过程中，误差总是客观存在的。进行化学实验之前，实验者有必要了解实验过程中，特别是物质组成测定过程中误差产生的原因及误差出现的规律。此外，为了准确表达具体的测定数值或测定结果，同时反映测量的精度，要掌握有效数字的概念。

1.4.1 测量中的误差

（1）误差的定义

误差是测量值与真实值的差值。但真实值通常是未知的，因此在实际工作中人们常用标准方法通过多次重复测定，将所测得的算术平均值作为真实值。误差的大小通常用绝对误差和相对误差来表示。

绝对误差可以表示为：

$$\text{绝对误差}(\Delta) = \text{测量值}(l) - \text{真实值}(x)$$

绝对误差是有量纲的量，其量纲与测量值和真值的量纲相同。

相对误差是绝对误差占真实值的百分数，其大小可以表示为：

$$\text{相对误差}(r) = \frac{\text{绝对误差}(\Delta)}{\text{真实值}(x)} \times 100\%$$

相对误差是量纲为 1 的量，常以百分数来表示。

（2）准确度与误差

准确度是指测量值与真实值之间的符合程度。准确度的高低常以误差的大小来衡量。即误差越小，准确度越高；误差越大，准确度越低。由于测量值（l）可能大于真实值（x），也可能小于真实值，所以绝对误差和相对误差都可能有正、有负。

【例 1-4】 若测量值为 20.30，真实值为 20.34，则

$$\text{绝对误差}(\Delta) = l - x = 20.30 - 20.34 = -0.04$$

$$\text{相对误差}(r) = \frac{\Delta}{x} \times 100\% = \frac{-0.04}{20.34} \times 100\% = -0.19\%$$

【例 1-5】 若测量值为 60.35，真实值为 60.39，则

$$\text{绝对误差}(\Delta) = l - x = 60.35 - 60.39 = -0.04$$

$$\text{相对误差}(r) = \frac{\Delta}{x} \times 100\% = \frac{-0.04}{60.39} \times 100\% = -0.06\%$$

上面两例中测定的绝对误差是相同的,但相对误差却相差很大,说明二者的含义是不同的,绝对误差表示的是测量值和真实值之差,而相对误差表示的是该误差在真实值中所占的百分率。

对于多次测量的数值,其准确度可按下式计算:

$$绝对误差(\Delta) = \frac{\sum x_i}{n} - x$$

式中 x_i——第 i 次测定的结果;
n——测定次数;
x——真实值。

$$相对误差(r) = \frac{\Delta}{x} \times 100\%$$

【例 1-6】 某样品测定 3 次结果分别为:0.1035g/L、0.1039g/L 和 0.1028g/L,标准样品含量为 0.1042g/L,求绝对误差和相对误差。

解:
$$平均值 = \frac{0.1035 + 0.1039 + 0.1028}{3} = 0.1034(g/L)$$

$$绝对误差(\Delta) = l - x = 0.1034 - 0.1042 = -0.0008(g/L)$$

$$相对误差(r) = \frac{\Delta}{x} \times 100\% = \frac{-0.0008}{0.1402} \times 100\% = -0.57\%$$

应注意的是有时为了表明一些仪器的测量准确度,用绝对误差来表示测量误差会更准确。例如分析天平的误差是 ±0.0002g,常量滴定管的读数误差是 ±0.01mL 等,都是用绝对误差来说明的。

(3) 精密度与偏差

由于误差的计算涉及被分析样品的真实值,而在定量分析前,被分析样品的真实值是未知的,因此常用被分析样品多次分析测定结果的平均值来代替真实值。这样,单次分析测定结果与多次分析测定结果平均值的差值称为偏差。精密度是指在相同条件下 n 次重复测定结果彼此相符合的程度。偏差的大小用精密度表示,偏差越小说明精密度越高。偏差的大小有以下几种表达方式。

① 绝对偏差 测定值与平均值之差。

$$绝对偏差(d_i) = x_i - \bar{x} \quad (i = 1, 2, \cdots, n)$$

式中 x_i——第 i 次测定值;
$\bar{x}$——n 次测定值的算术平均值,$\bar{x} = \frac{x_1 + x_2 + \cdots x_n}{n}$;
n——测定次数。

② 相对偏差

$$相对偏差 = \frac{d_i}{x} \quad (i = 1, 2, \cdots, n)$$

③ 平均偏差 绝对偏差的算术平均值。

$$平均偏差(\bar{d}) = \frac{|d_1| + |d_2| + \cdots + |d_n|}{n}$$

④ 相对平均偏差　多次分析测定结果的平均偏差与多次分析测定结果的平均值之比，通常用百分数表示：

$$相对平均偏差 = \frac{\bar{d}}{\bar{x}} \times 100\%$$

⑤ 标准偏差　标准偏差是偏差平方的统计平均值，又称均方根偏差，当测定次数 $n \to \infty$ 可表示为：

$$\sigma = \sqrt{\frac{\sum(x_i - \bar{x})^2}{n}} \quad (i = 1, 2, \cdots, n)$$

当 n 为有限次数时，即进行有限分析测定时的标准偏差用下式计算：

$$s = \sqrt{\frac{\sum(x_i - \bar{x})^2}{n-1}} \quad (i = 1, 2, \cdots, n)$$

【例 1-7】 计算下列一组测定值的平均偏差、相对平均偏差以及标准偏差。

66.71, 66.68, 66.75, 66.71, 66.69

解： 平均值 = (66.71 + 66.68 + 66.75 + 66.71 + 66.69)/5 = 66.71

$$平均偏差(\bar{d}) = \frac{|d_1| + |d_2| + \cdots + |d_n|}{n}$$

$$= \frac{|0| + |-0.03| + |0.04| + |0| + |-0.02|}{5} = 0.018$$

$$相对平均偏差 = \frac{\bar{d}}{\bar{x}} \times 100\% = \frac{0.018}{66.71} \times 100\% = 0.027\%$$

$$标准偏差(s) = \sqrt{\frac{\sum(x_i - \bar{x})^2}{n-1}} = \sqrt{\frac{(0)^2 + (-0.03)^2 + (0.04)^2 + (0)^2 + (-0.02)^2}{4}} = 0.027$$

（4）误差的来源

测量工作是在一定条件下进行的，外界环境、观测者技术水平和仪器本身构造的不完善等原因，都可能导致测量误差的产生。常把测量仪器、观测者的技术水平和外界环境三个方面综合起来，称为观测条件。观测条件不理想和不断变化，是产生测量误差的根本原因。具体来说，测量误差主要来自以下六个方面。

① 方法误差　方法误差又称理论误差，是由测定方法本身造成的误差，或是由于测定所依据的原理本身不完善而导致的误差。例如，在重量分析中，由于沉淀的溶解，共沉淀现象，灼烧时沉淀分解或挥发等；在滴定分析中，反应进行不完全或有副反应，干扰离子的影响，使滴定终点与化学计量点不能完全吻合，如此等等原因都会引起测定的方法误差。

② 仪器误差　仪器误差也称工具误差，是测定所用仪器不完善造成的。分析中所用的仪器主要指基准仪器（天平、玻璃量具）和测定仪器（如分光光度计等）。由于天平是分析测定中最基本的基准仪器，应由计量部门定期进行检校。市售的玻璃量具（容量瓶、移液管、滴定管、比色管等），其真实容量并非全部与其标称的容量相符，对一些要求较高的分析工作，要根据容许误差范围，对所用的仪器进行容量检定。分析所用的测定仪器，要按说明书进行调整。在使用过程中应随时进行检查，以免发生异常而造成仪器误差。

③ 人员误差　由于测定人员的分辨力、反应速度和固有习惯而引起的误差称为人员误

差。这类误差往往因人而异，因此可采取多人分析，以平均值报告分析结果的方法予以限制。

④ 环境误差　由测定环境所带来的误差叫做环境误差。例如室温、湿度不是所要求的标准条件，测定时仪器振动和电磁场、电网电压、电源频率等变化的影响，室内照明对于滴定终点判断的影响等。实验时如发现环境条件对测定结果有影响，应重新进行测定。

⑤ 随机误差　随机误差是在相同条件下，对同一量进行多次测定时，单次测定值与平均值之间差异的绝对值和符号无法预计的误差。这种误差是由测定过程中各种随机因素的共同影响而造成的。在一次测定中，随机误差的大小及其正负是无法预计的，没有任何规律性。在多次测定中，随机误差的出现具有统计规律性，即：随机误差有大有小，时正时负；绝对值小的误差比绝对值大的误差出现的次数多；在一定条件下得到的有限个测定值中，其误差的绝对值不会超过一定的界限；在测定的次数足够多时，绝对值相近的正误差与负误差出现的次数大致相等，此时正负误差相互抵消，随机误差的绝对值趋向于零。分析人员在用平均值报告分析结果时，正是运用了这一概率定律，在排除了系统误差的情况下，用增加测定次数的方法，使平均值成为与真实值较吻合的估计值。

⑥ 过失误差　这类误差明显地歪曲测定结果，是由测定过程中犯了不应有的错误而造成的。例如，标准溶液超过保存期，浓度或价态已经发生变化而仍在使用；器皿不清洁；不严格按照分析步骤或不准确地按分析方法进行操作；弄错试剂或吸管；试剂加入过量或不足；操作过程中试样受到大量损失或污染；仪器出现异常未被发现；读数、记录及计算错误等，都会产生误差。过失误差无一定的规律可循，这些误差基本上是可以避免的。消除过失误差的关键，在于分析人员必须养成专心、认真、细致的良好工作习惯，不断提高理论认识和操作技术水平。

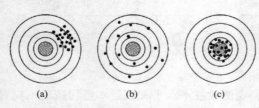

图1-1　精密度与准确度示意图

（5）准确度与精密度的关系

准确度和精密度是两个不同的概念，它们之间既有联系也有区别。以打靶为例（图1-1）：(a) 表示弹着点密集而离靶心（真实值）甚远，说明精密度高，随机误差小，但系统误差大，准确度低；(b) 表示精密度和准确度均较低，即随机误差和系统误差都大；(c) 表示系统误差与随机误差均小，准确度和精密度均较高。

测定的精密度高，测定结果也越接近真实值。但不能绝对地认为精密度高，准确度也高，因为系统误差的存在并不影响测定的精密度。相反，如果没有较好的精密度，获得较高准确度的可能性也很小。可以说精密度是保证准确度的先决条件。

（6）减小实验误差的方法

要提高分析结果的准确度，必须考虑在分析过程中可能产生的各种误差，采取有效措施，将这些误差减少到最小。

① 选择合适的分析方法　各种分析方法的准确度是不同的。化学分析法对高含量组分的测定能获得准确和较满意的结果，相对误差一般在千分之几。而对低含量组分的测定，化学分析法就达不到这个要求。仪器分析法虽然误差较大，但由于灵敏度高，可以测出低含量组分。在选择分析方法时，一定要根据组分含量及对准确度的要求，在可能条件下选择最佳分析方法。

② 增加平行测定次数　如前所述增加测定次数可以减少随机误差。在一般分析工作中，测定次数为 2～4 次。如果没有意外误差发生，基本上可以得到比较准确的分析结果。

③ 消除测定中的系统误差

a. 做空白实验。即在不加试样的情况下，按试样分析规程在同样操作条件下进行分析，所得结果的数值称为空白值。试样分析结束后，从试样结果中扣除空白值就得到比较可靠的分析结果。

b. 注意仪器校正。具有准确体积和质量的仪器，如滴定管、移液管、容量瓶和分析天平，都应进行校正，以消除仪器不准所引起的系统误差。

c. 作对照试验。对照试验就是用同样的分析方法在同样的条件下，用标样代替试样进行的平行测定。在分析过程中检查有无系统误差存在，做对照试验是最有效的办法。通过对照试验可以校正测试结果，消除系统误差。

1.4.2　实验结果的记录

（1）有效数字

科学实验要得到准确结果，不仅要求正确选用实验方法和实验仪器，而且要求正确记录实验数据。所谓正确记录是指正确记录数字的位数。在测量结果的数字表示中，若干位可靠数字和一位可疑数字，便构成了有效数字。

有效数字保留的位数，应根据分析方法与仪器的准确度来确定。

【例 1-8】　在分析天平上称取试样 0.6000g，这不仅表明试样的质量 0.6000g，还表明称量的误差在±0.0002g 以内。如将其质量记录成 0.60g，则表明该试样是在电子台秤上称量的，其称量误差为±0.02g，故记录数据的位数不能任意增加或减少。

【例 1-9】　在分析天平上测得称量瓶的质量为 8.3620g，这个记录说明有 5 位有效数字，最后一位是可疑的。分析天平只能称准到 0.0002g，即称量瓶的实际质量应为（8.3620±0.0002）g。无论计量仪器如何精密，其最后一位数字总是估计出来的。

对于滴定管、移液管和吸量管，它们都能准确测量溶液体积到 0.01mL。

【例 1-10】　当用 50mL 滴定管测定溶液体积时，如测量体积大于 10mL 小于 50mL 时，应记录为 4 位有效数字，写成 17.26mL；如测定体积小于 10mL，应记录 3 位有效数字，例如写成 6.78mL。

【例 1-11】　当用 25mL 移液管移取溶液时，应记录为 25.00mL；当用 5mL 吸量管取溶液时，应记录为 5.00mL。

【例 1-12】　当用 250mL 容量瓶配制溶液时，所配溶液体积即为 250.0mL。当用 50mL 容量瓶配制溶液时，应记录为 50.0mL。

从上述实例可见，有效数字和仪器的准确程度有关，即有效数字不仅表明数量的大小而且也反映测量的准确度。总而言之，测量结果所记录的数字，应与所用仪器测量的准确度相适应。

（2）有效数字的确定

有效数字的确定一方面要与测量仪器的精度相一致，另一方面还要考虑有效数字的运算要求。有效数字的计算，遵循"四舍六入五成双""先进舍，后运算"的原则，因此在计算前需按照以下"修约规则"对数字进行修约：

① 在加减计算中，各数所保留的小数点后的位数应与所给各数中小数点后位数最少的相同。例如：23.62，0.0083 和 1.643 三数相加时，首先根据取舍规则对数字进行修约，然后计算，则为 23.62+0.01+1.64=25.27。

② 在乘除计算中，应以有效数字最少或百分误差最大的数字为准，对其他各数按上述规则修约后，再进行计算。所得积或商的精度也不应大于相乘或相除各数值中精度最小数值的精度。例如：0.121，25.6432，1.0578 三数相乘时，将数字进行修约后，写成 0.121×25.6×1.06=3.28。

③ 在对数计算中，真数与对数的有效位数应相同。

④ 计算平均值时，若为四个和多于四个数平均，则平均数的有效位数可增加一位。

⑤ 对于 π、e 等常数，有效数字的位数可以任意确定。

1.4.3 实验数据处理

实验得到的数据往往较多，为了清晰明了地表示实验结果进而形象直观地对其进行分析，需要对实验数据进行处理。无机与分析化学的实验数据处理方法主要有列表法和作图法。

(1) 列表法

列表法简明紧凑、便于比较，是表达实验数据最常用的方法之一，也是本教材中主要采用的数据表示方法。将各种实验数据列入一种设计得体、形式紧凑的表格内，可起到化繁为简的作用，便于对获得的实验结果进行相互比较，有利于分析和阐明某些实验结果的规律性（表 1-1）。

表 1-1 数据记录表

试剂体积及体积比	1	2	3	4	5	6	7	8	9	10	11
$V_{NH_4Fe(SO_4)_2}$/mL	0.00	1.00	2.00	3.00	4.00	5.00	6.00	7.00	8.00	9.00	10.00
V_L/mL	10.00	9.00	8.00	7.00	6.00	5.00	4.00	3.00	2.00	1.00	0.00
体积比(Volume ratio)=$\dfrac{V_{(Fe^{3+})}}{V_{(Fe^{3+})}+V_{(L)}}$											
Abs											

使用列表法表示数据的方法如下：

① 表格名简洁准确并置于表格上方，同时将表格的顺序号放在表名前。

② 根据需要合理选择表中所列项目，包含实验所需必要信息。表中项目要包括名称和单位，并尽量采用符号表示。

③ 科研论文及文献中一般采用三线表，表中无竖线。表中数字写法整齐统一。同一竖行的数字，小数点要上下对齐。数字为零时，要保证有效数字的位数。如，有效位数为小数点后两位，则零应计为 0.00。

④ 表中的主项代表自变量，副项代表因变量。变量一般取整数或其他比较方便的数值，按递增或递减顺序排列。因变量的数值要注意有效位数的选择能够反映实验数据本身的误差。

⑤ 必要时可在表下加附注说明数据来源和表中无法反映的需要说明的其他问题。

（2）作图法

作图法形象直观，也是人们经常采用的一种数据表示方法。作图是将实验原始数据通过正确的作图方法画出合适的曲线（或直线），从而形象直观且准确地表现出实验数据的特点、相互关系和变化规律，如极大、极小和转折点等，并能够进一步求解，获得斜率、截距、外推值、内插值等（图1-2）。

作图法也存在作图误差，若要获得良好的图解效果，首先要获得高质量的图形。因此，作图技术直接影响实验结果的准确性。

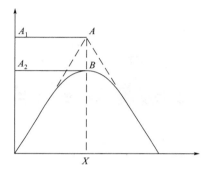

图1-2　浓比递变曲线

第二章
实验基本操作

2.1 基本器具

化学实验仪器大部分是玻璃制品,少部分为其他材质。玻璃制品具有较好的化学稳定性、良好的透明度,原料廉价易得且易于被加工成各种形状。表 2-1 是无机与分析化学实验中常见的器具名称、规格、用途及使用注意事项。

表 2-1 常用器具

仪器	规格	用途	注意事项
玻璃棒	玻璃制成的实心细棒	用于搅拌、引流等操作	①用过的玻璃棒用水洗涤后才能与另一种物质接触,以免污染试剂 ②转移液体时,应将盛放液体的容器口贴紧玻璃棒的下端靠在接受容器的内壁上,使液体沿玻璃棒缓缓流下
试管	玻璃或塑料制成	用作少量试剂的反应容器,常用于定性试验	①可直接用火加热,加热后不能骤冷 ②试管内盛放的液体量,不加热不超过 1/2,如加热不超过 1/3 ③加热试管内的固体物质时,管口应略向下倾斜以防凝结水回流至试管底部而使试管破裂
比色管	有开口和具塞两种,以最大容积表示,如 25mL、50mL	用于比较溶液颜色的深浅,快速定量分析中的目视比色	不能用试管刷刷洗,以免刮伤内壁,脏的比色管可以用铬酸洗液浸泡

续表

仪器	规格	用途	注意事项
锥形瓶	以容积表示,如 250mL、100mL	反应容器,因便于摇动,主要用于滴定操作	① 加热时,底部应垫石棉网 ② 可加热至较高温度,但温度不能变化过于剧烈,防止受热不均而破裂
碘量瓶	以容积表示,如 250mL	用于碘量法	注意保护磨口,以防产生漏隙
烧杯	以容积表示,规格较多,从 25mL 至 5000mL 不等	用于配制溶液、煮沸、蒸发、浓缩溶液及少量物质的制备等	① 加热时烧杯底要垫石棉网,防止受热不均而破裂 ② 所盛反应液体不得超过烧杯容量的 2/3,防止搅拌时液体溅出或沸腾时液体溢出
烧瓶	以容积表示,如 250mL、100mL、50mL 等	反应容器,反应物较多或需要较长时间加热时使用	可承受较高温度,加热时烧瓶底要垫石棉网,防止受热不均而破裂
量筒	以容积表示,如 200mL、100mL、50mL 等,上口大、下口小的称作量杯	用于量取一定体积的液体,在配制和量取浓度和体积不要求很精确的试剂时常用它来直接量取溶液	① 应竖直放置或持直,读数时视线与液面水平,读取与弯月面相切的刻度 ② 不可加热,不可用于溶解、稀释等操作,防止破裂 ③ 不可量热的液体
容量瓶	以容积表示,如 250mL、100mL、50mL 等 容量瓶有无色和棕色之分,棕色瓶用于配制需要避光的溶液	容量瓶用于配制体积要求准确的溶液,或作溶液的定量稀释	① 不能加热,不能在其中溶解固体 ② 瓶塞是磨口,不能互换以防漏水
称量瓶	称量瓶有高型和扁型两种	主要用于使用分析天平时称取一定量的试样	不能直接加热,瓶盖是磨口的不能互换

续表

仪器	规格	用途	注意事项
移液管/吸量管	移液管是中间有一膨大部分称为球部的玻璃管,球部上和下均为较细窄的颈,上端管刻有一条标线,亦称"单标线移液管" 吸量管是有刻度的玻璃管,用于移取非固定量的溶液	用于准确移取一定体积液体的量具	① 管口上无"吹"字样者,使用时末端的溶液不允许吹出 ② 移液管和吸量管均不能加热
滴定管	滴定管有常量和微量滴定管之分,常量滴定管有酸式和碱式两种。酸式滴定管用来盛强酸、氧化剂、还原剂等溶液,碱式滴定管用来盛碱溶液 滴定管有无色和棕色之分,无色的滴定管又有带蓝线和不带蓝线两种	滴定管是滴定时使用的精密仪器,用来测量自管内流出溶液的体积	① 注意量取溶液时应先排除滴定管尖端部分的气泡 ② 不能加热以及量取热的液体 ③ 酸碱滴定管不能互换使用
滴瓶	滴瓶有无色和棕色之分	用于盛装各种试剂,棕色瓶用于盛装需避光的试剂	① 注意滴管要专用不能"张冠李戴",不准乱放,不得弄脏,防止玷污试剂 ② 用滴管吸液时不能吸得太满,也不能平放或倒置防止试剂侵蚀橡胶头 ③ 滴加试剂时,滴管要垂直,使每次滴加试剂的量一样
洗瓶	塑料材质,多为 500mL	用于盛装去离子水或利用去离子水洗涤沉淀和容器时使用	不可装入自来水
漏斗	常见有 60°角短颈标准漏斗、60°角长颈标准漏斗	主要用于过滤操作和向小口容器倾倒液体	不能用火直接加热
分液漏斗	常见有球形、梨形和筒形等。	主要用于互不相溶的两种液体分层和分离 球形分液漏斗适用于萃取分离操作;梨形分液漏斗可用于分离互不相溶的液体;筒形在合成反应中常用来随时加入反应试液	① 不能加热 ② 玻璃活塞不能互换

续表

仪器	规格	用途	注意事项
玻璃砂芯漏斗	由烧结玻璃制成,也叫耐酸漏斗。根据其孔径大小,分成 G1 到 G6 六种规格	用于过滤酸液,用酸类处理	① 在加热或冷却时应注意缓慢进行 ② 使用时不宜过滤氢氟酸、热浓磷酸、热或冷的浓碱液
布氏漏斗/抽滤瓶	布氏漏斗以直径表示,如10cm、8cm 等 抽滤瓶以容积表示,如500mL、250mL 等	用于减压过滤	不能用火直接加热
试剂瓶	常见试剂瓶有细口试剂瓶、广口试剂瓶,试剂瓶有无色和棕色之分,又有磨口和非磨口之分	试剂瓶用于盛装各种试剂,广口试剂瓶常用于盛放固体药品,细口试剂瓶和滴瓶常用于盛放液体药品,棕色瓶用于盛装应避光试剂	① 非磨口试剂瓶用于盛装碱性溶液或浓盐溶液,使用橡胶塞或软木塞 ② 磨口试剂瓶盛装酸、非强碱性试剂或有机试剂,瓶塞不能调换以防漏气 ③ 若长期不用,应在瓶口和瓶塞间加放纸条便于开启 ④ 不能用火直接加热,不能在瓶内久贮浓碱、浓盐溶液
点滴板	瓷制,有白色和黑色两种,按凹穴数目分,有十二穴、九穴、六穴等	用于点滴反应,一般不需要分离的沉淀反应,尤其是显色反应	① 不能加热 ② 白色沉淀用黑板,有色沉淀用白板
酒精灯	以酒精为燃料	用于加热	① 酒精灯的灯芯要平整,如已烧焦或不平整,要用剪刀修正 ② 添加酒精时,不超过酒精灯容积的 2/3,不少于容积的 1/4 ③ 绝对禁止向燃着的酒精灯里添加酒精,以免失火 ④ 绝对禁止用酒精灯引燃另一只酒精灯,要用火柴点燃 ⑤ 用完酒精灯,必须用灯帽盖灭,不可吹灭 ⑥ 不要碰倒酒精灯,万一洒出的酒精在桌上燃烧起来,应立即用湿布或沙子扑盖

续表

仪器	规格	用途	注意事项
酒精喷灯	有座式酒精喷灯和挂式酒精喷灯两种。座式酒精喷灯的酒精贮存灯座内，挂式喷灯的酒精贮存罐悬挂于高处	用于高温加热，火焰温度可达100~800℃	① 严禁使用开焊的喷灯 ② 严禁用其他热源加热灯壶 ③ 若经过两次预热后，喷灯仍然不能点燃时，应暂时停止使用。检查接口处是否漏气，喷出口是否堵塞和灯芯是否完好，待修好后方可使用 ④ 喷灯连续使用时间以30~40min为宜 ⑤ 在使用中如发现灯壶底部凸起时应立刻停止使用，查找原因
表面皿	以直径表示，如15cm、12cm、9cm等	主要用作烧杯盖，防止灰尘落入和加热时液体进溅等	不能直接用火加热
蒸发皿	有平底和圆底两种形状，口大底浅蒸发速度快	主要用于液体蒸发	① 耐高温但不宜骤冷 ② 蒸发溶液时一般放在石棉网上加热，如液体量多可直接加热，但液体量以不超过深度的2/3为宜
坩埚	材料分瓷、石英、铁、银、镍、铂等，以容积表示，有50mL、40mL、30mL等	用于灼烧固体	① 灼烧时，放在泥三角上直接加热，不需石棉网 ② 取下的灼热坩埚不能直接放置在桌面上，而要放在石棉网上 ③ 灼热的坩埚不能骤冷
坩埚钳	材料为铁或铜，表面常镀镍、铬	夹坩埚加热或往高温电炉、马弗炉中放、取坩埚，亦可用于夹取热的蒸发皿	① 使用时必须用干净的坩埚钳，防止弄脏坩埚中药品。坩埚钳用后，应尖端向上平放在实验台上，如温度很高则应放在石棉网上防止烫坏实验台 ② 实验完毕，将钳子擦干净，放入实验柜中，干燥放置
石棉网	以石棉和铁丝作为材料，以铁丝网边长表示	加热时常垫在平底玻璃仪器与热源中间，由于石棉是一种不良导体，它能使受热物体均匀受热，不致造成局部高温	① 石棉脱落的不能使用 ② 不能与水接触，以免石棉脱落或铁丝锈蚀 ③ 不可卷折，石棉涂层松脆易损坏
泥三角	材料为铁丝和瓷管，有大小之分	用于放置加热的坩埚和小蒸发皿	① 灼热的泥三角不要滴冷水，以免瓷管破裂 ② 选择泥三角时，要使放置在上面的坩埚所露出的上部，不超过本身高度的1/3

续表

仪器	规格	用途	注意事项
三角架	材料为铁	放置较大或较重的加热容器,或与石棉网、铁架台等配合在一套实验装置中作支持物	① 放置除水浴锅外的其他加热容器,应垫以石棉网 ② 下面加热火焰的位置要合适,一般用氧化焰加热
干燥器	干燥器的中下部口径略小,上面放置带孔瓷板,瓷板上放置待干燥物品,瓷板下放有干燥剂 常用干燥剂有 P_2O_5、碱石灰、硅胶、$CaSO_4$、CaO、$CaCl_2$、$CuSO_4$、浓硫酸等。固态干燥剂可直接放在瓷板下面。液态干燥剂放在小烧杯中再放到瓷板下面 直径从 100mm 至 500mm 不等	主要用于保持固态、液态样品或产物的干燥,也用来存放防潮的小型贵重仪器和已经烘干的称量瓶、坩埚等	① 要沿边口涂抹一薄层凡士林研合均匀至透明,使顶盖与干燥器本身保持密合,不致漏气 ② 开启顶盖时,应稍用力使干燥器顶盖向水平方向缓缓错开,取下的顶盖应翻过来放稳 ③ 热的物体应冷却到接近室温时,再移入干燥器内 ④ 干燥器洗涤过后要吹干或风干,切勿用加热或烘干的方法去除水汽 ⑤ 久存的干燥器在室温低、顶盖打不开时,可用热毛巾捂热或暖风吹化开启
研钵	有玻璃研钵、瓷研钵、铁研钵和玛瑙研钵等	主要用于研磨固体物质	① 玻璃研钵、瓷研钵适用于研磨硬度较低的物料,硬度大的物料应用玛瑙研钵 ② 研钵不能用火直接加热,不能敲击,只能压或研碎 ③ 研磨物质的总量不宜超过研钵容积的1/3,易爆物质如 $KClO_3$ 等只能轻轻压碎不能研磨
蝴蝶夹、铁夹、铁圈、铁架台	由铁架、铁圈及铁夹等组合而成	用于固定或放置反应容器蝴蝶夹用于固定酸式滴定管和碱式滴定管	① 仪器固定在铁架台上时,仪器和铁架的重心应落在铁架台的底盘中间 ② 用铁夹夹持仪器时,应以仪器不能转动为宜,不能夹紧过松,过松易脱落,过紧可能夹破仪器 ③ 加热后的铁圈不能撞击或摔落在地

2.2 仪器的洗涤与干燥

2.2.1 玻璃仪器的洗涤

为得到准确实验结果,进行化学实验前首先要求把仪器洗涤干净,实验后也要立即对用过的仪器进行洗涤。

(1) 普通玻璃仪器

对试管、烧杯等普通玻璃仪器,可在容器内先注入1/3左右的自来水,选用大小合适的毛刷蘸取去污粉刷洗。用水冲洗后仪器内壁能均匀地被水润湿且不沾水珠,表明已基本洗干净。如有水珠沾仪器内壁，应重新洗涤以去除油污。用自来水洗净后的试管再用少量去离子水冲洗2～3次。

操作要点：使用毛刷洗涤试管时，注意刷子顶端的毛必须顺着伸入试管中，并用食指顶住试管底部，避免刷洗时用力过猛而将试管底部击穿。

（2）精确定量仪器

在使用精确定量仪器（如滴定管、移液管、容量瓶等）时，仪器的洁净程度要求较高，且其形状不规则，不宜用刷子刷洗，因此常用洗液（浓 H_2SO_4 和饱和 $K_2Cr_2O_7$ 溶液混合物）进行洗涤。方法是先将仪器用水冲洗，然后加入少量洗液，转动容器使其内壁全部为洗液浸润，经一段时间后，将洗液倒回原瓶，再用自来水冲洗，最后用去离子水冲洗2～3次。

操作要点：使用洗液前，先用水刷洗仪器，尽量除去其中污物；尽量把仪器中残留水倒掉，以免将洗液稀释，影响洗涤效果；洗液用后应倒回原瓶，以便重复使用；洗液具有极强的腐蚀性，易灼伤皮肤及衣物，使用时应注意安全；洗液变绿后不再具有氧化性和去污能力。

2.2.2 仪器内沉淀垢迹的洗涤

实验时，一些不溶于水的沉淀垢迹常牢固地沾在容器内壁，需根据其性质，选用适当的试剂，通过化学方法除去。常见垢迹的处理方法见表 2-2。

表 2-2 常见垢迹的处理方法

污垢痕迹	洗涤方法
沾在器壁上的 MnO_2、$Fe(OH)_3$、碱土金属的碳酸盐等	用盐酸处理，MnO_2 垢迹需浓度 6mol/L 以上的 HCl 才能洗掉
沉积在器壁上的铜和银	用硝酸处理
沉积在器壁上的难溶性银盐	一般用 $Na_2S_2O_3$ 洗涤。Ag_2S 垢迹用热浓 HNO_3 处理
沾在器壁上的硫黄	用煮沸的石灰水处理
残留在容器内的 Na_2SO_4 或 $NaHSO_4$ 固体	加水煮沸使其溶解，趁热倒掉
不溶于水、酸或碱的有机化合物和胶质等污迹	常用乙醇、丙酮、苯、四氯化碳、石油醚等有机溶剂洗
煤焦油污迹	用浓碱浸泡(约 24h)再用水冲洗
蒸发皿和坩埚内的污迹	一般可用浓 HNO_3 和王水洗涤
瓷研钵内的污迹	取少量食盐放在研钵内研洗，倒去食盐，再用水洗净

2.2.3 仪器的干燥

仪器干燥方法有多种，根据需求选择适宜干燥方法。

① 晾干：把洗净的仪器倒置于干净的仪器筐中或木钉上晾干。

② 烤干：用煤气灯或酒精灯小火烤干。

③ 吹干：用吹风器冷风吹干。

④ 烘干：将洗净的仪器放在电热鼓风干燥箱中或挂在鼓风干燥器（图 2-1）上烘干。

⑤ 有机溶剂快速干燥：先用少量易挥发有机溶剂淋洗一遍，

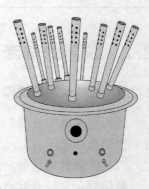

图 2-1 鼓风干燥器

然后晾干。

操作要点：带有刻度的容量仪器，如移液管、容量瓶、滴定管等不能用高温加热法干燥。

2.3 容量仪器的使用及溶液配制

定量分析中常用的玻璃量器（简称量器）有滴定管、移液管（吸量管）、容量瓶（量瓶）、量筒和量杯等。

2.3.1 量筒及其使用

量筒是最常用的度量液体体积仪器，有多种规格，按需选用。

操作要点：读取量筒中液体的体积数值时，视线需与量筒内液体弯月面的最低点保持水平（图2-2）。仰视结果偏低，俯视结果偏高。

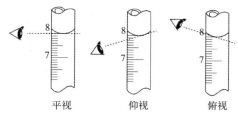

图2-2 量筒的读数方法

2.3.2 容量瓶及其使用

容量瓶是用来配制一定准确体积溶液或稀释溶液到一定浓度的容器。有多种规格，按需选用。当容量瓶内液体弯月面与容量瓶上端细颈处的刻度线相切时，瓶内液体体积可准确至±0.01mL。

操作要点：使用容量瓶前需先清洗再检漏。检漏的方法是将容量瓶盛约1/2体积水，盖上塞子，一手按住瓶塞，一手托住瓶底，倒置容量瓶。观察瓶塞周围有无漏水现象，再转动瓶塞180°，如仍不漏水，即可使用。

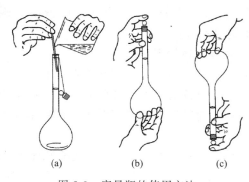

图2-3 容量瓶的使用方法

用固体配制溶液，需先在烧杯中用少量溶剂把固体溶解（必要时可加热）。待溶液冷至室温时，把溶液转移至容量瓶中［图2-3(a)］。然后用溶剂冲洗烧杯壁2~3次，冲洗液都移至容量瓶中，再加溶剂至容量瓶标线处。接近标线时，用滴管逐滴滴加溶剂至弯月面最低处恰好与标线相切。最后摇动容量瓶，使瓶中溶液混合均匀。摇动时，一手手指抵住瓶底边缘（不可用手心握住），一手按住瓶塞，将容量瓶倒置缓慢摇动［图2-3(b)(c)］，如此重复多次即可。

2.3.3 移液管及其使用

移液管是准确量取一定体积液体的仪器，有两种类型。一种为球形移液管，只有一条标线，只能用来移取一种体积的液体；另一种为刻度移液管（也称吸量管），管壁有多条刻度标线，可用来量取多种体积的溶液。

(1) 移液管的洗涤

使用移液管前应先对其认真洗涤。洗涤时，先用自来水洗涤2~3次，再用去离子水

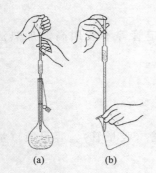

图 2-4 移液管的使用方法

洗涤 2~3 次，最后用待移液洗涤 2~3 次，每次洗涤的操作都相似。

操作要点：将移液管尖端部分插入液体[图 2-4(a)]，用洗耳球在移液管上端将少量液体吸起，然后用食指堵住移液管上端口，将移液管从液体中提出，慢慢将其一边旋转一边放平，让管内液体慢慢流淌致使整支移液管都被润湿，随后将液体从移液管中放出弃掉。如此重复多次，直至按规定将移液管洗涤干净。

(2) 移液管的使用

操作要点：移取液体时，将移液管尖端部分深深插入液体中，再用洗耳球在移液管上端将液体吸至高于刻度线后迅速用食指堵住移液管的上端口，将移液管提离液面后，使其垂直并微微移动食指，使液体弯月面恰好下降至与刻度线相切，然后用食指压紧管口，使液面不再下降。小心将移液管转入接收容器内，使管尖靠在接收容器的内壁，保持移液管垂直而接收容器倾斜，松开食指，让液体自由流出[图 2-4(b)]。移液管溶液只计算自由流出的液体，故留在管内的残留液不能吹出。使用完毕，将移液管清洗干净放在移液管架上。

2.3.4 滴定管及其使用

滴定管主要在容量分析中作滴定用，也可用于准确取液。滴定管有两种，一种是下端有玻璃活塞的酸式滴定管（简称酸管），另一种是下端有乳胶管和玻璃球代替活塞的碱式滴定管（简称碱管）。除碱性溶液用碱管盛装外，其他溶液一般都用酸管盛装。

① 检查　使用前应检查酸管的活塞是否配合紧密，碱管的乳胶管是否老化，玻璃球是否合适。如不合要求则进行更换。随后检查是否漏液：检查酸管时，关闭活塞，向管中注满水，直立静置 2min，仔细观察有无水滴漏出，然后将活塞旋转 180°，再直立观察 2min，观察是否有水漏出。检查碱管时，直立观察 2min 无水漏出即可。

若碱管漏液，可能是玻璃球过小或乳胶管老化，根据具体情况进行更换。

若酸管漏液或活塞转动不灵活，则应给活塞涂油，方法如下：将酸管平放在桌面上，取下活塞，先用滤纸吸干活塞、活塞槽上的水，把少许凡士林涂在活塞小孔的两侧，然后将活塞插入活塞槽中，按同一方向旋转活塞多次，直至观察活塞全部透明且不漏水。最后用乳胶圈将活塞固定，防止脱落破损。

② 洗涤　滴定管在使用前需进行洗涤，将滴定管内的污物洗涤干净后，用去离子水和盛装液各淌洗三次，每次应使滴定管的全部内壁和尖嘴玻管都得到淌洗，淌洗液的用量每次为 6~10mL。淌洗的目的是确保盛装溶液的浓度保持不变。

③ 装液　滴定管经淌洗后，从贮液瓶中直接倒入溶液至"0"刻度以上，观察下端尖嘴部位是否有空气泡，若有气泡，可倾斜滴定管 45°角，迅速旋转活塞，让溶液冲出将气泡带走。对碱管中的气泡，可将乳胶管向上弯曲，用手指挤压玻璃球稍上沿的乳胶管，让溶液冲出，气泡即被赶走（图 2-5）。

④ 读数　滴定前，将管内液面调节至 0.00~1.00mL 范围

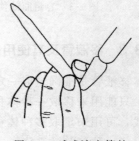

图 2-5 碱式滴定管的排气方法

内的某一刻度，等待 1~2min 若液面位置不变，则可读取滴定前管内液面位置的读数，滴定结束后，再读取管内液面位置的读数，两次读数之差，即为滴定所用溶液的体积。

读数时应注意：滴定管尖嘴处不应留有液滴，尖嘴内不应留有气泡；滴定管应保持垂直，为此，通常将滴定管从滴定管夹上取下用拇指和食指拿住滴定管上端无刻度处，在其自然下垂时读数；每次读数前应等待 1~2min 让附着在管内壁的溶液流下再读数。读数时，对无色或浅色溶液应读滴定管内液面弯月面最低处的位置，对深色溶液（如 $KMnO_4$、I_2 等），由于弯月面不清晰可读取液面最高点的位置。不论读什么位置，都应保持视线与应读位置处于同一水平线。否则读数会产生误差，如图 2-6 所示。

⑤ 滴定操作　将滴定管垂直固定于蝴蝶夹上，可在铁架台上放置白色衬底以便清楚地观察滴定过程中溶液颜色的变化。

操作酸管时，有刻度一面对着操作者，活塞柄在右方，由左手拇指、食指和中指配合动作，控制活塞旋转，无名指和小指向手心弯曲轻贴于尖脚端［图 2-7(a)］，旋转活塞时要轻轻向手心用力，以免活塞松动漏液。

操作碱管时，用左手拇指和食指在玻璃球右侧稍上处挤压乳胶管，使玻璃珠与乳胶管间形成一条缝隙，溶液即可流出［图 2-7(b)］。不要挤压玻璃珠下方的乳胶管，否则，气泡会进入玻璃尖嘴。

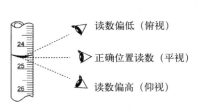

图 2-6　滴定管的读数

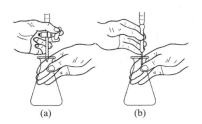

图 2-7　滴定管的使用方法

滴定操作在锥形瓶中进行，用右手拇指、食指和中指持锥形瓶，使瓶底距离滴定台 2~3cm，滴定管尖嘴伸入瓶内约 1cm，利用手腕的转动使锥形瓶旋转，左手按上述方法操作滴定管，一边滴加溶液一边转动锥形瓶。

操作要点：滴定时，左手不能离开活塞，注意观察滴落点周围溶液颜色的变化，以便控制溶液的滴速。滴定开始时，滴速可较快，使溶液一滴接一滴落下但不能成线流；接近终点时，要逐滴滴加并摇匀，最后以半滴加入，即控制溶液在尖嘴口悬而不落，用锥形瓶内壁粘下悬挂的液滴，再用洗瓶吹出少量去离子水冲洗锥形瓶内壁，摇匀。如此重复操作，直至达到滴定终点。滴定时，溶液可能由于摇动附到锥形瓶内壁的上部，故在接近终点时，用洗瓶吹出少量去离子水冲洗锥形瓶内壁。滴定结束，将管内溶液弃去，洗净滴定管倒置于蝴蝶夹上备用。

2.4　试剂及其取用

2.4.1　化学试剂的等级

常用化学试剂根据纯度不同一般分四个级别，见表 2-3。

表 2-3 试剂的规格与适用范围

级别	名称	代号	瓶标颜色	适用范围
一级	优级纯	GR	绿色	痕量分析和科学研究
二级	分析纯	AR	红色	一般定性定量分析实验
三级	化学纯	CP	蓝色	一般化学制备和教学实验
四级	实验试剂	LR	棕色或其他颜色	一般化学实验辅助试剂

除上述一般试剂外，还有一些特殊要求的试剂，如指示剂、生化试剂和超纯试剂（如电子纯、光谱纯、色谱纯）等，均会在试剂瓶标签上注明，使用时请注意。

固体试剂装于广口瓶中，液体试剂或配好的溶液盛放在细口瓶或带滴管的滴瓶中。见光易分解的试剂（如 $AgNO_3$ 等）盛放在棕色瓶内。试剂瓶上的标签应注明试剂名称、浓度、配制日期等信息。

2.4.2 试剂的取用

（1）固体试剂

操作要点：打开试剂瓶，将瓶塞倒置于实验台上或放在洁净的表面皿上，随后用清洁、干燥的药匙（塑料、玻璃或牛角质）取用。试剂取出后应立即盖紧瓶塞，将试剂瓶放回原处；已取出的试剂，不能再倒回试剂瓶内，多取的试剂可放在指定容器内供他人用。往试管特别是湿试管中加入固体试剂时，可将药匙伸入试管约 2/3 处，或将取出的试剂放在一张对折的纸条中再伸入试管，放入试管底部［如图 2-8(a)(b)］。块状固体则应沿管壁慢慢滑下。

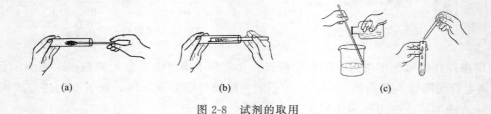

图 2-8 试剂的取用

（2）液体试剂

操作要点：

① 将细口瓶瓶塞取下，倒置于实验台上或放置在洁净的表面皿中，一手拿住容器，一手握住试剂瓶，试剂瓶的标签朝向手心，倒出所需量的试剂。倒完后，将试剂瓶口在容器上靠一下，再使瓶身垂直，以免液滴沿外壁流下。将液体从试剂瓶中倒入烧杯中时，亦可用一手握住试剂瓶，一手拿玻璃棒，使棒的下端斜靠在烧杯中，将瓶口靠在玻棒上，让液体沿玻棒向下流［如图 2-8(c)］。

② 从滴瓶中取用少量试剂时，提起滴管，使管口离开液面，用手指捏紧滴管上部的橡皮头排去空气，再把滴管伸入试剂瓶中吸取试剂。往试管中加试剂时，只能把滴管头放在试管口上方，严禁将滴管伸入试管内。滴瓶上的滴管不能混用，防止试剂污染。不能用实验者的滴管吸取试剂，以免污染试剂。不得将吸有试剂的滴管横置或滴管口向上斜放，以免液体流入橡皮头内，使试剂受污染或橡皮头被腐蚀。

2.5 加热方法

实验室中,常使用酒精灯、酒精喷灯、电炉等进行加热。

2.5.1 灯的使用

(1) 酒精灯的使用

酒精灯一般用玻璃制成,其灯罩带有磨口,不用时,必须将灯罩罩上,以免酒精挥发。酒精易燃,使用酒精灯的注意事项见表2-1。

(2) 酒精喷灯的使用

酒精喷灯一般由金属制成,使用前,先在预热盆中注入酒精至满,然后点燃盆内的酒精,以加热铜质灯管,等盆内酒精将近燃完时,开启开关。此时由于酒精在灼热灯管内汽化,并与表面气孔的空气混合,用火柴在管口点燃,可获得较高温度。调节开关的螺丝,可以控制火焰的大小,用毕后,向右旋紧开关,即可使灯焰熄灭。

应注意,在开启开关、点燃火焰之前,必须充分灼烧灯管;否则,酒精在管内不会全部汽化,会有液体酒精由管内喷出,形成"火雨",甚至引起火灾。在这种情况下,必须赶快熄灭喷灯,待稍冷后再往预热盆中添满酒精,重新预热灯管。喷灯不用时,必须关好储罐的开关,以免酒精漏失,造成危险。

2.5.2 加热方法

实验中常用的加热器具有烧杯、烧瓶、锥形瓶、蒸发皿、坩埚、试管等,这些玻璃仪器能够承受一定的温度变化,但不能骤热或骤冷,因此在加热前,必须将容器外面的水擦干,加热后不能立即与潮湿的物体接触。

(1) 在试管中加热

① 加热液体 试管中的液体一般可直接放在火焰上加热,但易分解的物质应在水浴中加热。

操作要点:用试管夹夹持试管中上部,试管稍微向上倾斜[图2-9(a)]。先加热液体中上部,再慢慢向下移动。加热过程中不时移动试管,使液体各部分受热均匀,防止蒸汽骤然产生,液体冲出管外。试管口不要对着别人和自己,以免溶液溅出时把人烫伤或腐蚀损伤。离心试管应在水浴中加热。

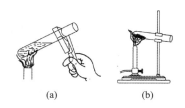

图2-9 加热试管中的液体和固体

② 加热固体 操作要点:将试管(硬质玻璃试管)固定在铁架台上或用试管夹夹住,试管口稍微向下倾斜[图2-9(b)]。加热开始时来回移动试管,使试管均匀受热,随后集中在固体部位加热。加热结束,待试管自然冷却后进行洗涤。

(2) 在水浴中加热

要加热在100℃时易分解的溶液,或需保持一定温度进行实验时,需用水浴加热[图2-10(a)]。水浴锅一般是铜和铝制成的,上面放置大小不同的圆环,以承受不同大小的器皿

（必要时可用放置在石棉网上盛水的大烧杯来代替水浴锅）。

操作要点：水浴锅内盛水量不要超过总容量的 2/3，并应随时补充少量的热水，以保持其中有占容量 2/3 的水量；当不慎将水浴锅中的水烧干时，应立即停止加热，待水浴锅冷却后，再加水继续使用。若需较严格地控制水浴温度，应选用电热型水浴装置［图2-10(b)］。

(a) 水浴加热　　　　(b) 恒温水浴锅　　　　(c) 沙浴加热

图 2-10　不同加热装置

（3）在油浴或沙浴中加热

当被加热的物质要求受热均匀，且温度又高于 100℃ 时，可使用沙浴或油浴。沙浴是一个盛有均匀细沙的铁盘［图2-10(c)］。

操作要点：加热时，被加热器皿的下部埋在沙中，若要测量沙浴温度，可将温度计插入沙中。用油代替水浴锅中的水，即为油浴。加热浓缩液体或加热固体时，要进行充分搅拌，使液体或固体受热均匀。

（4）高温加热

当需要在高温下加热固体时，可把固体放在坩埚内，用氧化焰加热［图2-11(a)］。

操作要点：先用小火使坩埚均匀受热，随后用大火燃烧加热。夹取高温下的坩埚时，需用干净的坩埚钳，且把坩埚钳的尖端先放在火焰上预热，再去夹取。使用完坩埚钳后，尖端向上平放，如温度过高则平放在石棉网上。切勿将还原焰接触坩埚底部，以免在坩埚底部集结炭黑，以致坩埚破裂。还可用管式炉［图2-11(b)］和马弗炉［图2-11(c)］等电器进行加热。

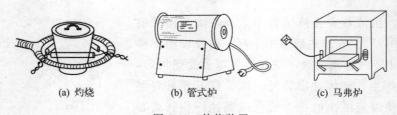

(a) 灼烧　　　　(b) 管式炉　　　　(c) 马弗炉

图 2-11　灼烧装置

2.6　蒸发、浓缩与结晶

2.6.1　溶液的蒸发和浓缩

当物质的溶解度很大，溶液浓度很低时，要使溶质结晶析出必须通过加热使溶液蒸发。蒸发到一定程度后冷却，即可析出晶体。

操作要点：溶液的蒸发和浓缩一般在蒸发皿中进行，对热稳定的溶液，可直接加热蒸发，但易分解的溶液则需在水浴上加热蒸发。溶液的体积不超过蒸发皿容积的 2/3；溶液不宜剧烈地沸腾，否则容易溅出；在沸水浴上蒸发溶液时，需随时向水浴锅中加水以免把水烧干；不要使热蒸发皿骤冷，以免炸裂。

2.6.2 结晶

结晶是指当溶质超过其溶解度时，晶体从溶液中析出的过程。当溶液蒸发或浓缩到一定浓度或过饱和时，如将溶液冷却或加入几粒晶体或搅动溶液，都能使晶体析出。

操作要点：从溶液中析出晶体的颗粒大小与结晶条件有关。如溶液浓度高，溶质溶解度小，冷却速度快，析出的晶体细小，相反则得到较大颗粒的晶体。形成晶体颗粒较大时，母液中的杂质容易被包裹在晶体的内部，降低晶体纯度。如将溶液迅速冷却并加以搅拌，则得到的晶体颗粒较细，但纯度较高。

2.6.3 晶体的干燥

晶体干燥的方法主要包括：自然风干法、加热风干法、真空干燥法、吸附剂法和微波干燥法。

操作要点：根据晶体的性质、干燥的要求和设备条件选择合适的干燥方法。同时注意控制干燥温度、时间和环境条件，避免对晶体结构和性质产生不利影响。

2.7 固液分离

固体和液体的分离，常用倾析法、过滤法和离心分离法。

（1）倾析法

当沉淀的密度或结晶颗粒较大，静置后容易沉降至容器底部时，常用倾析法分离。

操作要点：将沉淀上部清液缓慢倾入另一容器内，使沉淀和溶液分离，如图 2-12 所示。洗涤沉淀时，可向沉淀中加入少量洗涤剂充分搅拌后将沉淀静置，让沉淀沉降，再小心地倾析出洗涤液。如此重复 2～3 次，即可把沉淀洗净。

图 2-12　倾析法过滤

（2）过滤法

常见的过滤方法有三种，即常压过滤、减压过滤和热过滤。

① 常压过滤　在常压下用普通漏斗过滤的方法称为常压过滤法。当沉淀物为胶体或微细的晶体时，用此法过滤较好，但过滤速度较慢。

操作要点：过滤时，先取一圆形滤纸对折两次，拨开一层即为 60°的圆锥形（若用方形滤纸，则对折两次后剪成扇形），如图 2-13(a) 所示。将其放入漏斗中。用少量去离子水润湿滤纸，再用食指或玻棒挤压滤纸四周，挤出滤纸与漏斗之间的气泡，使滤纸紧贴漏斗壁上。漏斗中滤纸的上边缘应低于漏斗边缘。过滤时，把漏斗放在漏斗架或铁架台上，调整漏斗架高度，使漏斗尖端紧靠在接受容器的内壁，以使滤液能顺器壁流下，不致四溅［图 2-13(b)］。用倾析法将溶液沿玻棒于三层滤纸处缓缓倾入漏斗中。漏斗中液面高度应低于滤纸上沿 2~3mm，如果沉淀需要洗涤可在溶液转移完后，往盛沉淀的容器中加入少量洗涤剂，充分搅拌并放置，待沉淀下沉后再把洗涤剂倾入漏斗，如此重复 2~3 次，再把沉淀转移到滤纸上。

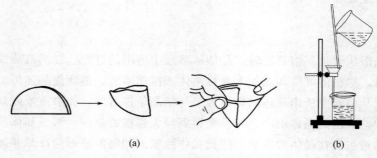

图 2-13　滤纸的折叠和常压过滤

② 减压过滤　减压可加速过滤，并使沉淀抽得比较干，但不宜用于过滤颗粒太小的沉淀和胶状沉淀。减压过滤的仪器装置如图 2-14 所示。

操作要点：剪好一张比布氏漏斗内径略小的圆形滤纸，大小以能覆盖布氏漏斗所有小孔为准。将滤纸平整地放在漏斗中，用少量去离子水润湿滤纸，把漏斗插入单孔橡胶塞内并塞在吸滤瓶上，注意漏斗下端的斜削面要对着吸滤瓶侧面的支管。用橡皮管把吸滤瓶与抽气泵相连，打开抽气泵即可抽滤。抽滤时，先用倾析法，加入量不要超过漏斗高度的 2/3，先将上部溶液沿玻棒倒入漏斗中，然后再将沉淀移入漏斗中滤纸的中间部分。过滤时，不能让滤液上升到吸滤瓶支管的水平位置，否则滤液将被抽出吸滤瓶。在抽滤过程中，不得突然关闭抽气泵。如欲取出滤液或停止抽滤，应先将吸滤瓶支管的橡皮管取下，再关上抽气泵，否则水会倒灌进入抽滤瓶。

过滤完后，先将吸滤瓶支管的橡皮管拆下，关闭泵，取下漏斗，使漏斗颈口朝上，轻轻敲打漏斗边缘，使滤纸和沉淀脱离漏斗而进入接受容器。滤液则从吸滤瓶的上口倾出，不要从侧面的尖嘴倒出，以免弄脏滤液。

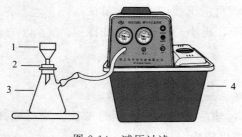

图 2-14　减压过滤
1—布氏漏斗；2—橡胶塞；3—吸滤瓶；4—抽气泵

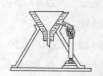

图 2-15　热过滤

③ 热过滤　如果某些溶液中的溶质在浓度高时易结晶析出，我们又不希望这些溶质在过滤的过程中析出而留在滤纸上，这就需要进行热过滤。

操作要点：热过滤时，把玻璃漏斗放在铜质的热漏斗内，热漏斗中装有热水，并用酒精灯加热（图 2-15）以维持滤液温度。也可在过滤时把玻璃漏斗放在水浴锅上用水蒸气进行加热，此法较为简便，可代替热漏斗。热过滤选用漏斗的颈部越短越好，以免过滤时溶液在漏斗颈内停留过久，因散热降温而析出晶体造成堵塞。

（3）离心分离法

离心分离法的工作原理是选用离心机（图 2-16）在高速旋转时产生离心力，将试管中的沉淀迅速聚集于试管底部。

操作要点：将盛有悬浊液的离心管放入离心机管套内，在与之对称的另一套管内装入相同质量的离心管，使离心机保持平衡。接通电源，设定转速和离心时间，开始离心。离心结束后，待离心机自然停下后，打开上盖，从管套中取出离心管，倾去上清液，留下沉淀。如需洗涤沉淀，可向沉淀中加入少量洗涤剂和沉淀剂，充分摇匀后再次离心，弃去溶液。

(a) 小型离心机　　　(b) 低速离心机　　　(c) 高速离心机

图 2-16　离心机

2.8　试纸的使用

试纸的作用是通过其颜色变化来测试溶液的性质，主要用于定性或定量分析，其特点是操作简便、快速，并具有一定的精确度。目前无机与分析化学实验中常用的试纸包括以下几种。

2.8.1　pH 试纸

pH 试纸用于检测溶液的酸碱性，不同 pH 值的溶液可使试纸呈现不同的颜色。常用的 pH 试纸有广泛 pH 试纸和精密 pH 试纸。广泛 pH 试纸可粗略地测量溶液的 pH 值，测量范围 pH=1~14。精密 pH 试纸的测量精确度较高，测量范围较窄。

操作要点：将一小块 pH 试纸放在点滴板或白瓷板上，用玻棒沾一点待测溶液并与 pH 试纸接触（不能把试纸扔进待测液中），试纸被待测溶液润湿变色，随后尽快与标准色阶板比较，确定 pH 值或 pH 范围。

2.8.2　碘化钾-淀粉试纸

碘化钾-淀粉试纸用于定性检验一些氧化性气体，如 Cl_2、SO_3 等。

操作要点：用去离子水将试纸润湿后卷在玻璃棒顶端放在试管口，如待测的氧化性气体逸出，就会使试纸中的 I^- 氧化成 I_2，I_2 与淀粉作用，试纸变为蓝紫色。使用时，注意不能让试纸长时间与氧化性气体接触。特别是气体氧化性强、浓度大时要更加注意，因为 I_2 有可能进一步被氧化为 IO_3^- 而使试纸褪色，影响测定结果。

2.8.3 醋酸铅试纸

醋酸铅 $Pb(Ac)_2$ 试纸用于定性检查 H_2S 气体，使用方法与碘化钾-淀粉试纸相同。如反应中有 H_2S 产生，则试纸因生成 PbS 而呈褐色或亮灰色。

2.8.4 石蕊试纸

石蕊试纸用于检验气体或溶液的酸碱性，通常有红色石蕊试纸和蓝色石蕊试纸两种。使用方法与碘化钾-淀粉试纸相同。若有酸性气体产生，可使蓝色石蕊试纸变红色；若有碱性气体产生，可使红色石蕊试纸变为蓝色。

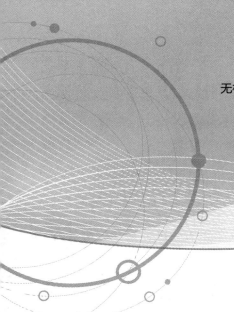

第三章
基础实验

实验一 氯化钠的提纯

一、实验目的

1. 掌握提纯 NaCl 的原理和方法。
2. 学习减压过滤、蒸发浓缩、结晶等基本操作。
3. 了解 Ca^{2+}、Mg^{2+}、SO_4^{2-} 等离子的定性鉴定。

二、实验原理

化学试剂或医药用 NaCl 都是以粗食盐为原料提纯而得。粗食盐中含有 Ca^{2+}、Mg^{2+}、K^+ 和 SO_4^{2-} 等可溶性杂质和泥沙等不溶性杂质。选择适当的试剂可使 Ca^{2+}、Mg^{2+}、SO_4^{2-} 等离子形成难溶盐沉淀而除去,一般先在食盐溶液中加 $BaCl_2$ 溶液,除去 SO_4^{2-}:

$$Ba^{2+} + SO_4^{2-} = BaSO_4 \downarrow$$

然后再在溶液中加 Na_2CO_3 溶液,除 Ca^{2+}、Mg^{2+} 和过量的 Ba^{2+}:

$$Ca^{2+} + CO_3^{2-} = CaCO_3 \downarrow$$
$$Ba^{2+} + CO_3^{2-} = BaCO_3 \downarrow$$
$$4Mg^{2+} + 2H_2O + 5CO_3^{2-} = Mg(OH)_2 \cdot 3MgCO_3 \downarrow + 2HCO_3^-$$

过量的 Na_2CO_3 溶液用 HCl 中和,粗食盐中的 K^+ 仍留在溶液中。由于 KCl 溶解度比 NaCl 大,且在粗食盐中含量较少,所以在蒸发和浓缩食盐溶液时,NaCl 先结晶析出,而 KCl 仍留在溶液中。

三、仪器与药品

仪器:分析天平,酒精灯,布氏漏斗,抽滤瓶,蒸发皿,玻璃棒,石棉网,100mL 烧杯。

药品：NaCl（粗）；Na_2CO_3（饱和溶液）；HCl（6mol/L）；$(NH_4)_2C_2O_4$（饱和溶液）；$BaCl_2$（1mol/L）；NaOH（6mol/L）；HAc（2mol/L）；镁试剂Ⅰ；pH试纸。

四、实验步骤

1. 粗食盐的溶解

称取3.0g粗食盐于100mL烧杯中，加入30mL水，加热（酒精灯）搅拌使其溶解（不溶性杂质沉于底部）。

2. 除去SO_4^{2-}

加热溶液至近沸，边搅拌边滴加1mol/L $BaCl_2$溶液至SO_4^{2-}沉淀完全，继续加热5min，减压过滤，弃去沉淀。

3. 除Ca^{2+}、Mg^{2+}和过量的Ba^{2+}等阳离子

将上步得到的滤液加热至近沸，边搅拌边滴加饱和Na_2CO_3溶液，直至不再产生沉淀为止，继续加热5min后，减压过滤，弃去沉淀。

4. 除去剩余的CO_3^{2-}

向溶液中滴加6mol/L HCl，加热搅拌，中和到溶液pH值为2～3。

5. 浓缩与结晶

将溶液倒入蒸发皿中蒸发浓缩至表面出现一层NaCl晶膜，冷却，抽滤。将NaCl晶体转移至蒸发皿中，在石棉网上小火烘干。冷却后称量，计算产率。

6. 产品纯度的检验

取粗食盐和提纯后的产品NaCl各1.0g，分别溶于5mL去离子水中，然后进行下列离子的定性检验。

① SO_4^{2-}的检验　向两支试管中分别加入上述粗、纯NaCl溶液约1mL，分别加入2滴6mol/L HCl和2滴1mol/L $BaCl_2$溶液，比较两溶液中沉淀产生的情况。

② Ca^{2+}检验　向两支试管中分别加入粗、纯NaCl溶液约1mL，加2mol/L HAc使其呈酸性，再分别加入3～4滴饱和$(NH_4)_2C_2O_4$溶液，如有白色CaC_2O_4沉淀产生，表示有Ca^{2+}存在。比较两溶液中沉淀产生的情况。

③ Mg^{2+}检验　向两支试管中分别加入粗、纯NaCl溶液约1mL，先各加入约5滴6mol/L NaOH，摇匀，再分别加入2滴镁试剂Ⅰ溶液，溶液若有蓝色絮状沉淀产生，表明有镁离子存在。观察两溶液的颜色。

五、注意事项

1. 蒸发浓缩时，不能将溶液蒸干。
2. 浓缩液自然冷却至室温。

六、思考题

1. 在除去Ca^{2+}、Mg^{2+}、SO_4^{2-}时，为何先加$BaCl_2$溶液，然后再加Na_2CO_3溶液？
2. 能否用$CaCl_2$代替毒性大的$BaCl_2$来除去食盐中的SO_4^{2-}？
3. 在除Ca^{2+}、Mg^{2+}、SO_4^{2-}等杂质离子时，能否用其他可溶性碳酸盐代替Na_2CO_3？
4. 能否用重结晶的办法提纯氯化钠？

Experiment 1 Purification of Sodium Chloride

1. Purpose

(1) To learn the principle and method of purifying sodium chloride

(2) To learn the operation of vacuum filtration, evaporation, concentration and crystallization

(3) To learn the methods of qualitative test for Ca^{2+}, Mg^{2+} and SO_4^{2-}

2. Principle

Sodium chloride, which is used as a chemical or medical reagent, is purified from crude salt. There are not only insoluble impurities in the crude salt, such as sediment, but also soluble impurities, such as Ca^{2+}, Mg^{2+}, K^+ and SO_4^{2-}. To remove Ca^{2+}, Mg^{2+} and SO_4^{2-}, add appropriate reagents to produce insoluble precipitates.

First, add $BaCl_2$ to the crude salt solution to remove SO_4^{2-}.

$$Ba^{2+} + SO_4^{2-} =\!=\!= BaSO_4 \downarrow$$

Then add Na_2CO_3 to remove Ca^{2+}, Mg^{2+} and excessive Ba^{2+}.

$$Ca^{2+} + CO_3^{2-} =\!=\!= CaCO_3 \downarrow$$
$$Ba^{2+} + CO_3^{2-} =\!=\!= BaCO_3 \downarrow$$
$$4Mg^{2+} + 2H_2O + 5CO_3^{2-} =\!=\!= Mg(OH)_2 \cdot 3MgCO_3 \downarrow + 2HCO_3^-$$

The excessive Na_2CO_3 can be neutralized with HCl. The low content soluble impurity KCl, having a different solubility from sodium chloride, can be removed by recrystallization. It will be retained in the solution when NaCl crystals form.

3. Apparatus and Chemicals

Apparatus: Analytical balance; alcohol burner; buchner funnel; filter flask; evaporating dish; 100mL beaker.

Chemicals: NaCl (crude salt); HCl (6mol/L); HAc (2mol/L); NaOH (6mol/L); $BaCl_2$ (1mol/L); Na_2CO_3 (saturated); $(NH_4)_2C_2O_4$ (saturated); magnesium reagent I; pH test paper.

4. Procedure

(1) Dissolving crude salt

Weigh 3.0g of crude salt in a 100mL beaker, add 30mL of water, heat (alcohol burner) and stir to make it dissolve.

(2) Removing SO_4^{2-}

Heat the solution to boiling, and then add 1mol/L $BaCl_2$ solution while stirring until the precipitation is complete. After continuing to boil the mixture for 5 minutes, vacuum filter the mixture.

(3) Removing Mg^{2+}, Ca^{2+} and excess of Ba^{2+}

Heat the above filtrate to boil. Add saturated Na_2CO_3 solution while stirring until the precipitation is complete. Vacuum filter the mixture, and discard the precipitates.

(4) Removing residuary CO_3^{2-}

Heat and stir the solution, add 6mol/L HCl until the pH value of the solution is about 2-3.

(5) Concentration and crystallization

Pour the solution into an evaporating dish and evaporate until a layer of NaCl crystal film appears on the surface. Cool and filter. Transfer the NaCl crystal to an evaporating dish and dry it over low heat on an asbestos mesh. After cooling, weigh and calculate the yield.

(6) Product purity analysis

Take crude salt and purified product NaCl each 1.0g, add about 5mL distilled water. Then perform the following qualitative analyses.

① SO_4^{2-} In two test tubes, transfer 1mL of crude salt solution and 1mL of purified salt solution, respectively. In each test tube, add two drops of 6mol/L HCl and two drops of 1mol/L $BaCl_2$. Compare the precipitates in the two test tubes.

② Ca^{2+} In two test tubes, transfer 1mL of crude salt solution and 1mL of purified salt solution, respectively. In each test tube, add 2mol/L HAc to acidulate, and 3-4 drops of saturated $(NH_4)_2C_2O_4$. If white calcium oxalate precipitates, it indicates the presence of calcium ions. Compare the white precipitates (CaC_2O_4) in the two test tubes.

③ Mg^{2+} In two test tubes, transfer 1mL of crude salt solution and 1mL of purified salt solution, respectively. In each test tube, add five drops of 6mol/L NaOH and 2 drops of magnesium Ⅰ reagent. The blue precipitates confirm the presence of Mg^{2+}. Compare the blue precipitates in the two test tubes.

5. Notes

(1) When being evaporated and concentrated, the solution cannot be dried up.

(2) The concentrate should be naturally cooled to room temperature.

6. Questions

(1) When removing Mg^{2+}, Ca^{2+} and SO_4^{2-}, why add $BaCl_2$ first and then Na_2CO_3?

(2) Can we substitute calcium chloride by toxicant barium chloride to remove the SO_4^{2-} in salt?

(3) Can we use another soluble carbonate to replace Na_2CO_3 in order to remove Mg^{2+}, Ca^{2+} and Ba^{2+}?

(4) Can we use the method of recrystallization to purity sodium chloride?

实验二 硝酸钾的制备和提纯

一、实验目的

1. 利用物质溶解度随温度变化的差别，学习用转化法制备硝酸钾晶体。
2. 练习间接热浴和重结晶操作。

二、实验原理

工业上常采用转化法制备硝酸钾晶体,其反应如下:

$$NaNO_3 + KCl \rightleftharpoons NaCl + KNO_3$$

NaCl 的溶解度随温度变化不大,在较高温度时,其溶解度最小。而 $NaNO_3$、KCl 和 KNO_3 在高温时具有较大的溶解度且温度降低时溶解度明显减小(表 3-1)。根据这几种盐溶解度的差异,将一定浓度的 $NaNO_3$ 和 KCl 混合液加热浓缩,当温度逐渐升高时,由于 KNO_3 溶解度增加很多,达不到饱和,不析出;而 NaCl 的溶解度增加甚少,随浓缩、溶剂的减少,NaCl 析出。通过热过滤除去 NaCl,将此溶液冷却至室温,即有大量 KNO_3 析出,而 NaCl 仅有少量析出,从而得到 KNO_3 粗产品。再经过重结晶提纯,可得到纯品。

表 3-1 四种盐在不同温度下的溶解度

盐	溶解度(100g H_2O)/g							
	0℃	10℃	20℃	30℃	40℃	60℃	80℃	100℃
KNO_3	13.3	20.9	31.6	45.8	63.9	110.0	169	246
$NaNO_3$	73	80	88	96	104	124	148	180
KCl	27.6	31.0	34.0	37.0	40.0	45.5	51.1	56.7
NaCl	35.7	35.8	36.0	36.3	36.6	37.3	38.4	39.8

三、仪器与药品

仪器:分析天平;水浴锅;布氏漏斗;抽滤瓶;100mL 烧杯;小试管。

药品:硝酸钠(工业级);氯化钾(工业级);$AgNO_3$(0.1mol/L);硝酸(5mol/L);氯化钠标准溶液。

四、实验步骤

1. 硝酸钾的制备

称取 10g $NaNO_3$ 和 8.5g KCl,放入 100mL 烧杯中,随后加入 30mL H_2O,加热使固体溶解。待盐全部溶解后,继续加热,并不断搅拌,使溶液蒸发至原有体积的 2/3 左右。此时烧杯中有晶体析出。趁热过滤。滤液盛于小烧杯中自然冷却至室温。随着温度的下降,即有结晶析出。减压过滤,抽干。粗产品水浴烘干后称重。计算理论产量和产率。

2. 粗产品重结晶

除保留少量(0.1~0.2g)粗产品供纯度检验外,按粗产品:水=2:1(质量比)的比例,将粗产品溶于去离子水中。加热、搅拌、待晶体全部溶解后停止加热。若溶液沸腾时,晶体还未全部溶解,可再加极少量去离子水使其溶解。待溶液冷却至室温析出晶体后减压过滤,水浴烘干,得到纯度较高的硝酸钾晶体,称重。

3. 产品纯度定性检验

分别取 0.1g 粗产品和一次重结晶得到的产品放入两支小试管中,各加入 2mL 去离子水配成溶液。在溶液中分别加入 1 滴 5mol/L HNO_3 酸化,再各滴入 2 滴 0.1mol/L $AgNO_3$ 溶液,观察现象,进行对比,重结晶后的产品溶液应为澄清。否则应再次重结晶,直至合格。

五、注意事项

1. 冷却结晶时，不要骤冷，以防晶体过于细小。
2. 反应混合物一定要趁热快速减压抽滤，布氏漏斗在沸水中或烘箱中预热。

六、思考题

1. 制备硝酸钾晶体时，为何要将溶液进行加热和热过滤？
2. 硝酸钾中含有氯化钾和硝酸钠时，应如何提纯？
3. 何为重结晶？重结晶过程中需要注意哪些问题？

Experiment 2　Preparation and Purification of Potassium Nitrate

1. Purpose

（1）To learn how to prepare potassium nitrate crystals using the conversion method by utilizing the difference in solubility of substances with temperature changes.

（2）To practice indirect hot bath and recrystallization operations.

2. Principle

Conversion method is always used to prepare potassium nitrate in industry, and the reaction is as follows:

$$NaNO_3 + KCl \rightleftharpoons NaCl + KNO_3$$

The solubility of NaCl changes little with temperature and it is the least soluble at high temperature among these four salts. $NaNO_3$, KCl and KNO_3 have larger solubility at high temperature and their solubilities decrease significantly with temperature decreases (Table 3-1). According to the difference of the solubility of these four salts, mix a certain concentration of $NaNO_3$ and KCl solution, concentrate it by heating until the temperature reaches 118-120 ℃. KNO_3 can't precipitate because of its high solubility, while NaCl can. Remove NaCl by hot filtered, cool the solution to room temperature, a large amount of KNO_3 and only a small amount of NaCl precipitate. This is the crude KNO_3 product. After recrystallization, the pure KNO_3 product can be obtained.

Table 3-1　Solubility of four salts at different temperatures.

Salts	Solubility(in 100g H_2O)/g							
	0℃	10℃	20℃	30℃	40℃	60℃	80℃	100℃
KNO_3	13.3	20.9	31.6	45.8	63.9	110.0	169	246
$NaNO_3$	73	80	88	96	104	124	148	180
KCl	27.6	31.0	34.0	37.0	40.0	45.5	51.1	56.7
NaCl	35.7	35.8	36.0	36.3	36.6	37.3	38.4	39.8

3. Apparatus and Chemicals

Apparatus: Analytical balance; water bath; buchner funnel; filter flask; 100mL beaker; small test tube.

Chemicals: Sodium nitrate (industrial grade); potassium chloride (industrial grade); $AgNO_3$ (0.1mol/L); nitric acid (5mol/L).

4. Procedure

(1) Preparation of potassium nitrate

Weigh 10g of $NaNO_3$ and 8.5g of KCl into a 100mL beaker, then add 30mL of H_2O and heat to dissolve the solid. After all the salts are dissolved, continue heating and stirring continuously to evaporate the solution to about 2/3 of its original volume. At this point, crystals are precipitated in the beaker. Filter while hot. The filtrate is placed in a small beaker and naturally cooled to room temperature. As the temperature decreases, crystals precipitate. Use reduce-pressure filtration and drain. Weigh the crude product after drying in a water bath. Calculate the theoretical yield and yield.

(2) The recrystallization of crude KNO_3

Except for retaining a small amount (0.1-0.2g) of crude product for purity testing, dissolve the crude product in deionized water at a ratio of crude product to water = 2 : 1 (mass ratio). Heat, stir, and stop heating after all the crystals have dissolved. If the crystal has not completely dissolved when the solution boils, an additional small amount of deionized water can be added to dissolve it. Wait for the solution to cool to room temperature and precipitate crystals, then filter under reduced pressure and dry in a water bath to obtain high-purity potassium nitrate crystals. Weigh it and calculate the yield.

(3) Product purity analysis

Take 0.1g of crude product and the product obtained from one recrystallization respectively and place them into two small test tubes. Add 2mL of deionized water to each tube to prepare a solution. Add 1 drop of 5mol/L HNO_3 to the solution for acidification, and then add 2 drops of 0.1mol/L $AgNO_3$ solution to each solution. Observe the phenomenon and compare. The recrystallized product solution should be clear. Otherwise, it should be recrystallized again until qualified.

5. Notes

(1) When cooling crystals, do not cool them suddenly to prevent them from becoming too small.

(2) The reaction mixture must be rapidly depressurized and filtered while hot, and the buchner funnel needs to be preheated in boiling water or an oven.

6. Questions

(1) Why do we need to heat and filter the solution when preparing potassium nitrate

crystals?

(2) How should potassium nitrate be purified when it contains potassium chloride and sodium nitrate?

(3) What is recrystallization? What issues should be noted during the recrystallization process?

实验三　醋酸标准电离常数和电离度的测定

一、实验目的

1. 掌握醋酸标准电离常数和电离度的测定方法。
2. 加深对弱电解质电离平衡的理解。
3. 学习酸度计和容量瓶的使用方法。

二、实验原理

醋酸是弱电解质，在溶液中存在如下的电离平衡：
$$HAc \rightleftharpoons H^+ + Ac^-$$

其电离常数 $K_a^\ominus$ 的表达式为：
$$K_a^\ominus = \frac{([H^+]/c^\ominus)([Ac^-]/c^\ominus)}{[HAc]/c^\ominus} \tag{3-1}$$

在浓度足够大的 HAc 溶液中，设醋酸的起始浓度为 c_0，忽略水的电离，则平衡时 $[H^+]=[Ac^-]$，代入式（3-1）可得：
$$K_a^\ominus = \frac{[H^+]^2}{c_0 - [H^+]} \tag{3-2}$$

当 $c_0/K_a^\ominus \geqslant 500$ 时，式（3-2）简化为
$$K_a^\ominus = \frac{[H^+]^2}{c_0} \tag{3-3}$$

另外，HAc 的电离度 α 可表示为
$$\alpha = [H^+]/c_0 \tag{3-4}$$

如溶液中还有 NaAc，则因为同离子效应，HAc 电离度会降低，但 $K_a^\ominus$ 保持不变，此时计算平衡常数的公式为：
$$K_a^\ominus = \frac{[H^+]c_{NaAc}}{c_0} \tag{3-5}$$

在一定温度下，用酸度计测定已知浓度醋酸的 pH 值，根据 $pH = -\lg[H^+]$，换算出 $[H^+]$，代入式（3-3）、式（3-4）和式（3-5）中，可求该温度下醋酸的电离常数和电离度。

三、仪器与药品

仪器：酸度计；50mL 容量瓶；25mL 移液管；5mL 移液管；50mL 烧杯。
药品：HAc（0.10mol/L）；NaAc（0.10mol/L）。

四、实验步骤

1. 配制不同浓度的醋酸溶液

用移液管分别移取 5.00mL、10.00mL、25.00mL 已知浓度的 HAc 溶液于 3 个 50mL

容量瓶中，用去离子水稀释至刻度，摇匀。连同未稀释的 HAc 溶液，可得四种不同浓度的溶液，由稀到浓依次编号为 1、2、3、4。

用移液管分别移取 25.00mL HAc 和 5.00mL NaAc 溶液加入 50mL 容量瓶，用去离子水稀释至刻度，摇匀，编号为 5。

2. HAc 溶液的 pH 测定

取上述 5 种溶液各 20mL 分别加入 5 只干燥洁净的 50mL 烧杯中，使用酸度计测定其 pH 值。

五、数据记录及结果处理

将上述实验数据及数据处理记录到表 3-2 中。

表 3-2　实验数据及数据处理 [室温（℃）＝　　　　　]

序号	c/(mol/L)	pH	$[H^+]$/(mol/L)	$[Ac^-]$/(mol/L)	$K_a^\ominus$	α	t/℃
1							
2							
3							
4							
5							

六、注意事项

1. 5 个烧杯必须干燥洁净。
2. 测定上述 5 种醋酸溶液的 pH 值时，一定要按照由稀到浓顺序操作。

七、思考题

1. 实验中所用烧杯是否需要用 HAc 溶液润洗？
2. 如果改变所测 HAc 的温度，则电离度和标准电离常数有无变化？

Experiment 3　Determination of Ionization Constant of Acetic Acid and Its Ionization Degree

1. Purpose

(1) To master the method of measuring the standard ionization constant and ionization degree of acetic acid.

(2) To Deepen the understanding of weak electrolyte ionization equilibrium.

(3) To learn how to use acidity meters and volumetric flasks.

2. Principle

Acetic acid is a weak electrolyte, and there is a dissociation equilibrium in the solution as follows：

$$HAc \rightleftharpoons H^+ + Ac^-$$

The ionization constant of acetic acid $K_a^\ominus$ is expressed as:

$$K_a^\ominus = \frac{([H^+]/c^\ominus)([Ac^-]/c^\ominus)}{[HAc]/c^\ominus} \tag{3-1}$$

In an HAc solution with a large enough concentration, let the initial concentration of acetic acid be c_0, and ignore the ionization of water, then $[H^+]=[Ac^-]$ at equilibrium, equation (3-1) can be written as

$$K_a^\ominus = \frac{[H^+]^2}{c_0 - [H^+]} \tag{3-2}$$

When $c_0/K_a^\ominus \geqslant 500$, eq. (3-2) was simplified as

$$K_a^\ominus = \frac{[H^+]^2}{c_0} \tag{3-3}$$

In addition, the dissociation degree of HAc can be expressed as

$$\alpha = [H^+]/c_0 \tag{3-4}$$

If there is NaAc in the solution, the dissociation degree of HAc will decrease due to the same ion effect, but $K_a^\ominus$ remains unchanged. At this time, the formula for calculating the equilibrium constant is:

$$K_a^\ominus = \frac{[H^+] c_{NaAc}}{c_0} \tag{3-5}$$

At a certain temperature, the pH value of acetic acid with known concentration is determined by acidity meter. According to $pH = -\lg[H^+]$, $[H^+]$ is converted, and the ionization constant and dissociation of acetic acid at this temperature can be obtained by substituting it into equations (3-3), (3-4) and (3-5).

3. Apparatus and Chemicals

Apparatus: pH meter; 50mL volumetric flask; 25mL pipette; 50mL beaker.
Chemicals: HAc (0.10mol/L); NaAc (0.10mol/L).

4. Procedure

(1) Prepare acetic acid solution with various concentrations

Transfer 5.00mL, 10.00mL, 25.00mL HAc solution with known concentration into three 50mL volumetric bottles with deionized water, and shake well. Together with the undiluted HAc solution, four different concentrations can be obtained, and numbered 1, 2, 3, and 4 from dilute to concentrated.

Transfer 25.00mL HAc and 5.00mL NaAc solution into 50mL volumetric bottle with pipette, dilute to scale with deionized water, shake well, number it as 5.

(2) Determine the pH value of acetic acid solution

Take 20mL of each 5 solutions and add them to 5 dry 50mL beakers respectively. Determine their pH value by a pH meter. Calculate their degree of ionization and ionization constant.

5. Data Analysis

Table 3-2 lists the data and analysis results.

Table 3-2 The degree of ionization and ionization constant of HAc [r. t(℃) =   ]

No.	c/(mol/L)	pH	$[H^+]$/(mol/L)	$[Ac^-]$/(mol/L)	$K_a^\ominus$	α	t/℃
1							
2							
3							
4							
5							

6. Notes

(1) The five beakers must be clean and dry.

(2) When measuring the pH values of the above 5 acetic acid solutions, it is necessary to follow the order of dilution to concentration.

7. Questions

(1) Do the beakers used in the experiment need to be rinsed with HAc solution?

(2) If the temperature of the measured HAc is changed, do the ionization degree and standard ionization constant change?

实验四　磺基水杨酸合铁（Ⅲ）配合物组成及稳定常数的测定

一、实验目的

1. 掌握分光光度法测定溶液中配合物组成及稳定常数的原理和方法。
2. 学习使用分光光度计。

二、实验原理

磺基水杨酸与 Fe^{3+} 可以形成稳定的配合物。配合物的组成和颜色随溶液 pH 的不同而改变。当 pH<4 时，形成 1∶1 的紫红色配合物；pH＝4～9 时生成 1∶2 的红色配合物；pH＝9～11 时，生成 1∶3 的黄色配合物。本实验是在 pH≈2 时利用分光光度法测定磺基水杨酸合铁配合物的组成及稳定常数。反应方程式如下：

$$Fe^{3+} + \underset{\underset{COOH}{|}}{\underset{|}{SO_3^-}}\!\!-\!\!\!\bigcirc\!\!\!-\!\!OH \rightleftharpoons \left[SO_3^-\!\!-\!\!\bigcirc\!\!\!<\!\!\!\!{O-\atop C-O}\!\!\!>\!\!Fe^{3+}\right] + 2H^+$$

当一束具有一定波长的单色光通过一定厚度的有色溶液时，有色物质吸收了一部分光能，使透射光的强度（I_1）与入射光（I_0）相比有所减弱。对光的吸收程度和透过程度，通常用吸光度（A）和透光率（T）来表示。透光率是透射光强度与入射光强度之比，即：

$$T = \frac{I_1}{I_0}$$

吸光度是透光率的负对数，即：

$$A = -\lg T = -\lg \frac{I_1}{I_0}$$

吸光度越大说明入射光被有色物质吸收的程度越大，反之，吸光度越小，表示有色物质对光的吸收程度越小。吸光度 A 与有色物质的浓度、液层厚度成正比：

$$A = \xi bc$$

这就是朗伯-比尔定律。ξ 为摩尔吸光系数，当波长一定时，它是有色物质的特征常数。b 为液层厚度，当其不变时，吸光度 A 与有色物质的浓度 c 成正比。

本实验采用等摩尔系列法（也叫浓比递变法）测定磺基水杨酸与 Fe^{3+} 形成配合物的组成和稳定常数。即保持溶液中金属离子 M 和配体 L 总物质的量不变，而 M 和 L 的摩尔分数连续变化，配置成一系列溶液，测定其吸光度。在这一系列溶液中，有一些溶液的金属离子是过量的，而另一些的配体是过量的，这两部分溶液中的配离子浓度都不可能达到最大值，只有当溶液中金属离子和配体的摩尔比与配离子的组成一致时，配离子浓度最大，吸光度也最大。若以吸光度 A 为纵坐标、溶液中配体的摩尔分数为横坐标作图（图3-1），所得曲线出现一个高峰，它所对应的吸光度为 A_2。如果延长曲线两侧的直线线段部分，相交于点 A，此点所对应的吸光度为 A_1，即为吸光度的极大值。这两点所对应的配体的摩尔分数即为配离子的组成。如此点所对应的配体的摩尔分数为0.5，则中心离子的摩尔分数为0.5，即金属离子和配体之比是1∶1，该配离子（或配合物）的组成为 ML 型。

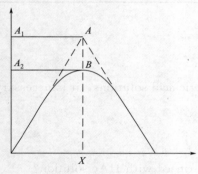

图3-1　浓比递变曲线

由于配离子（或配合物）有一部分电离，其浓度比未电离时的稍小，吸光度最大处所实际测得的最大吸光度 A_2 小于由曲线两侧延长所得点处即组成全部为 ML 配合物的吸光度 A_1。

因而配离子的电离度 $\alpha = \dfrac{A_1 - A_2}{A_1}$

ML 型配离子的稳定常数 $K_f^{\ominus}$ 与电离度 α 的关系如下：

$$K_f^{\ominus} = \frac{c_{eq(ML)}/c^{\ominus}}{[c_{eq(M)}/c^{\ominus}][c_{eq(L)}/c^{\ominus}]} = \frac{1-\alpha}{c\alpha^2} c^{\ominus}$$

式中，c 为 B 点 ML 配合物的浓度。

三、仪器与药品

仪器：分光光度计；100mL 容量瓶；25mL 移液管；25mL 烧杯；pH 试纸。

药品：$NH_4Fe(SO_4)_2$（0.01mol/L）；磺基水杨酸（0.01mol/L）；NaOH（6mol/L）；浓 H_2SO_4。

四、实验步骤

1. $NH_4Fe(SO_4)_2$ 溶液和磺基水杨酸溶液的配置

用移液管分别移取 10.00mL 0.01mol/L $NH_4Fe(SO_4)_2$ 溶液和磺基水杨酸溶液,置于两只 100mL 容量瓶中,用去离子水稀释,在稀释接近刻度线时,测定其 pH,若 pH 偏离 2,可通过滴加 1 滴浓 H_2SO_4 或 1 滴 6mol/L NaOH 溶液调整溶液 pH 为 2,随后滴加去离子水至刻度线并摇匀。

2. 制备等物质的量的磺基水杨酸合铁(Ⅲ)配合物

依表 3-3 所示溶液体积,依次在 11 个 25mL 烧杯中混合配制好等物质的量系列溶液。

表 3-3 数据记录表

试剂参数	1	2	3	4	5	6	7	8	9	10	11
$V_{NH_4Fe(SO_4)_2}$/mL	0.00	1.00	2.00	3.00	4.00	5.00	6.00	7.00	8.00	9.00	10.00
V_L/mL	10.00	9.00	8.00	7.00	6.00	5.00	4.00	3.00	2.00	1.00	0.00
体积比 $\dfrac{V_{(Fe^{3+})}}{V_{(Fe^{3+})}+V_L}$											
Abs											

3. 测定上述溶液的吸光度

利用分光光度计,在 $\lambda=500$nm,$b=1$cm 的条件下,以去离子水为空白,测定等物质的量磺基水杨酸合铁(Ⅲ)配合物溶液的吸光度 A,并记录。

五、结果处理

以体积比 $\dfrac{V_{(Fe^{3+})}}{V_{(Fe^{3+})}+V_L}$ 为横坐标,对应的吸光度 A 为纵坐标作图,从图 3-1 中找出最大的吸光度 A_1,计算在本实验条件下,磺基水杨酸合铁(Ⅲ)配合物的组成(n)和稳定常数($K_f^{\ominus}$)。

六、思考题

1. 实验中每个溶液的 pH 值是否相同?如不同对结果是否有影响?
2. 请说明 $n_M : n_L = 1 : 2$ 及 $n_M : n_L = 1 : 3$ 时吸光度的最大位置?

Experiment 4 Determination of Composition and Stability Constant of Sulfosalicylic Acid Iron (Ⅲ) Complex

1. Purpose

(1) To master the principle and method of determining the composition and stability constant of complexes in solutions using spectrophotometry.

(2) To learn to use a spectrophotometer.

2. Principle

Sulfosalicylic acid can form stable complexes with Fe^{3+}. The composition and color of the complex vary with the pH of the solution. When pH<4, a 1∶1 purple red complex is formed; generate a 1∶2 red complex at pH 4-9; when pH range from 9 to 11, a 1∶3 yellow complex is generated. This experiment uses spectrophotometry to determine the composition and stability constant of sulfosalicylic acid iron complex at pH≈2. The reaction equation is as follows:

$$Fe^{3+} + SO_3^-\text{-C}_6H_3\text{(OH)(COOH)} \rightleftharpoons [SO_3^-\text{-C}_6H_3\text{(O-Fe}^{3+}\text{)(C(O)-O)}] + 2H^+$$

When a monochromatic light with a certain wavelength passes through a colored solution of a certain thickness, the colored substance absorbs some of the light energy, causing the intensity of the transmitted light (I_1) to weaken compared to the incident light (I_0). The degrees of absorption and transmission of light are usually represented by absorbance (A) and transmittance (T). Transmittance is the ratio of transmitted light intensity to incident light intensity, i.e.,

$$T = \frac{I_1}{I_0}$$

Absorbance is the negative logarithm of transmittance, i.e.,

$$A = -\lg T = -\lg \frac{I_1}{I_0}$$

The higher the absorbance, the greater the degree to which the incident light is absorbed by colored substances. Conversely, the lower the absorbance, the less light is absorbed by colored substances. The absorbance A is directly proportional to the concentration of colored substances and the thickness of the liquid layer.

$$A = \xi bc$$

This is Lambert-Beer's law. The molar absorption coefficient is a characteristic constant of colored substances when the wavelength is constant. b is the thickness of the liquid layer, and when it remains constant, the absorbance A is directly proportional to the concentration c of the colored substance.

In this experiment, the equimolar series method (also known as concentration gradient method) is used to determine the composition and stability constant of the complex formed by sulfosalicylic acid and Fe^{3+}. The total amount of metal ion M and ligand L in the solution kept constant, while the molar fractions of M and L continuously change, a series of solutions are prepared and their absorbance is measured. In this series of solutions, some solutions have excess metal ions, while others have excess ligands. The concentra-

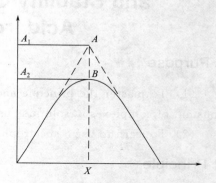

Fig 3-1 Concentration gradient curve

tion of coordination ions in these two parts of the solution cannot reach its maximum value. Only when the molar ratio of metal ions and ligands in the solution is consistent with the composition of coordination ions, can the concentration of coordination ions and the absorbance be maximized. If the absorbance A is plotted as the vertical axis and the molar fraction of ligands in the solution as the horizontal axis, the resulting curve shows a peak, corresponding to an absorbance of A_2. If the straight line segments on both sides of the extended curve intersect at point A, the absorbance corresponding to this point is A_1, which is the maximum absorbance. The molar fraction of the ligand corresponding to these two points is the composition of the coordination ion. If the molar fraction of the ligand corresponding to this point is 0.5, then the molar fraction of the central ion is 0.5, that is, the ratio of metal ions to ligands is 1 : 1, and the composition of the ligand ion (or complex) is ML type.

Some of the ions (or complexes) dissociate, and their concentration is slightly lower than that when they are not dissociated. The A_2 actually measured at the location of maximum absorbance is lower than the absorbance A_1 formed by extending the points on both sides of the curve, which is all ML complexes.

Therefore, the dissociation degree of coordination ions $\alpha = \dfrac{A_1 - A_2}{A_1}$.

The relationship between the stability constant $K_f^\ominus$ and dissociation degree α of ML type coordination ions is as follows:

$$K_f^\ominus = \frac{c_{eq(ML)}/c^\ominus}{[c_{eq(M)}/c^\ominus][c_{eq(L)}/c^\ominus]} = \frac{1-\alpha}{c\alpha^2} c^\ominus$$

Where, c represents the concentration of ML at point B.

3. Apparatus and Chemicals

Apparatus: spectrophotometer; 100mL volumetric flask; 25mL pipette; 25mL beaker; pH test paper.

Chemicals: $NH_4Fe(SO_4)_2$ (0.01mol/L); sulfosalicylic acid (0.01mol/L); NaOH (6mol/L); concentrated sulfuric acid.

4. Procedure

(1) Prepare $NH_4Fe(SO_4)_2$ and sulfosalicylic acid solution

Take 10.00mL of 0.01mol/L $NH_4Fe(SO_4)_2$ solution and sulfosalicylic acid solution separately in two 100mL volumetric flasks, dilute with deionized water, and measure their pH when the dilution is close to the mark. If the pH deviates from 2, adjust the solution pH to 2 by adding 1 drop of concentrated H_2SO_4 or 6mol/L NaOH solution, then add deionized water to the mark and shake well.

(2) Preparation of iron (Ⅲ) complexes with sulfosalicylic acid in equal amounts

Mix and prepare a series of equal substance solutions in 11 25-mL beakers according to the volume of the solution shown in Table 3-3.

Table 3-3 Experimental data

Reagents parameter	1	2	3	4	5	6	7	8	9	10	11
$V_{NH_4Fe(SO_4)_2}$/mL	0.00	1.00	2.00	3.00	4.00	5.00	6.00	7.00	8.00	9.00	10.00
V_L/mL	10.00	9.00	8.00	7.00	6.00	5.00	4.00	3.00	2.00	1.00	0.00
Volume ratio = $\dfrac{V_{(Fe^{3+})}}{V_{(Fe^{3+})}+V_L}$											
Abs											

(3) Measure the absorbance of the above solution

Using a spectrophotometer, measure and record the absorbance A of the equivalent amount of sulfosalicylic acid iron (Ⅲ) complex solution under the conditions of $\lambda=500$nm in a cuvette with $b=1$cm. Deionized water is used as the blank.

5. Data Analysis

Draw a graph with the volume ratio $\dfrac{V_{(Fe^{3+})}}{V_{(Fe^{3+})}+V_L}$ as the horizontal axis and the corresponding absorbance A as the vertical axis. Find the maximum absorbance A_1 from the graph and calculate the composition (n) and stability constant ($K_f^\ominus$) of the sulfosalicylic acid iron (Ⅲ) complex under the experimental conditions.

6. Questions

(1) Is the pH value of each solution the same in the experiment? Does the difference have an impact on the results?

(2) Please specify the maximum position of absorbance when $n_M:n_L=1:2$ and $n_M:n_L=1:3$.

实验五 BaSO₄溶度积常数的测定（电导率法）

一、实验目的

1. 学习沉淀的生成、陈化、离心分离、洗涤等基本操作。
2. 掌握难溶电解质溶度积测定的方法。
3. 学习电导率仪的使用方法。

二、实验原理

在难溶电解质$BaSO_4$的饱和溶液中，存在下列平衡：

$$BaSO_4 \rightleftharpoons Ba^{2+}+SO_4^{2-}$$

其溶度积为：$K_{sp(BaSO_4)}^\ominus = \left\{\dfrac{[Ba^{2+}]}{c^\ominus}\right\}\left\{\dfrac{[SO_4^{2-}]}{c^\ominus}\right\}$

由于难溶电解质的溶解度很小，很难直接测定。但只要有溶解作用，溶液中就有电离出来的带电离子，因此可以通过测定溶液的电导或电导率，再根据其与浓度的关系，计算出难溶电解质的溶解度，从而换算出溶度积。

电解质溶液导电能力的大小，可以用电阻 R 或电导 G（西门子，S）来表示，两者互为倒数：

$$G = \frac{1}{R}$$

因：
$$R = \rho \frac{l}{A}$$

所以：
$$G = \frac{1}{\rho} \cdot \frac{A}{l} = \kappa \cdot \frac{A}{l}$$

式中，ρ 为电阻率；l 为电阻长度；A 为电阻横截面积；κ 为电导率，表示在相距 1m、面积为 $1m^2$ 的两个电极之间溶液的电导，单位是西门子/米（S/m）。

摩尔电导率（Λ_m）是含 1mol 电解质的溶液置于相距为 1m 电极之间的电导。

$$\Lambda_m = \frac{\kappa}{c}$$

式中，Λ_m 的单位为 $S \cdot m^2/mol$。

极限摩尔电导率 Λ_∞ 表示在无限稀释情况下的摩尔电导率，是一个常数。因 $BaSO_4$ 的溶解度极小，因此其溶液可视为无限稀释溶液，存在以下关系式：

$$\Lambda_{mBaSO_4} = \Lambda_{mBa^{2+}} + \Lambda_{mSO_4^{2-}}$$

$$\Lambda_{\infty BaSO_4} = \Lambda_{mBaSO_4} \qquad c_{BaSO_4} = \frac{\kappa_{BaSO_4}}{1000\Lambda_{\infty BaSO_4}}$$

因：
$$c_{BaSO_4} = c_{Ba^{2+}} = c_{SO_4^{2-}}$$

$$K_{sp}^{\ominus} = [Ba^{2+}][SO_4^{2-}] = c_{BaSO_4}^2 = \left(\frac{\kappa_{BaSO_4}}{1000\Lambda_{\infty BaSO_4}}\right)^2 = \left[\frac{\kappa_{BaSO_4(溶液)} - \kappa_{H_2O}}{1000\Lambda_{\infty BaSO_4}}\right]^2$$

$$\Lambda_{\infty BaSO_4} = 2.872 \times 10^{-2} (S \cdot m^2)/mol$$

三、仪器与药品

仪器：电导率仪；离心机；酒精灯；100mL 烧杯；表面皿；玻璃棒。

药品：H_2SO_4（0.05mol/L）；$BaCl_2$（0.05mol/L）；$AgNO_3$（0.01mol/L）。

四、实验步骤

1. $BaSO_4$ 沉淀的制备

用量筒量取 30mL 0.05mol/L H_2SO_4 于 100mL 小烧杯中，将 H_2SO_4 溶液用酒精灯加热接近沸腾，边搅拌边滴加 0.05mol/L $BaCl_2$ 溶液 30mL，滴加完后，用 5mL 去离子水洗涤盛有 $BaCl_2$ 溶液的烧杯，并全部倒入上述混合液中，盖上表面皿。继续加热煮沸 5min，小火保温 10min，取下静置、陈化 15～20min，倾去上层清液。将沉淀和少量余液用玻璃棒搅成乳状，分次转移至离心管中，进行离心分离（3000r/min，5min），弃去上层溶液。

2. $BaSO_4$ 饱和溶液的配制

将 40mL 去离子水加入小烧杯中加热近沸腾，用于洗涤 $BaSO_4$ 沉淀。洗涤时每次加入 5mL 热去离子水，用玻璃棒将沉淀充分搅混再离心分离，弃去洗涤液。洗涤三次后，将洗涤液倒入干净烧杯中，用 0.01mol/L $AgNO_3$ 溶液验证洗涤液无 Cl^-（如产生沉淀，则需要继续洗涤,直至向洗涤液中滴加 $AgNO_3$ 溶液无沉淀产生)。往 $BaSO_4$ 沉淀中加入少量去离子水用玻璃棒搅匀转移至 50mL 已测电导率去离子水的小烧杯中,用表面皿覆盖,加热煮沸 3~5min,并不断搅拌 5min,静置,冷却,将上清液转移至烧杯中用于电导率的测定。

3. 电导率的测定

分别测定去离子水及 $BaSO_4$ 饱和溶液的电导率(κ,$\mu S/cm$),计算 $BaSO_4$ 的 $K_{sp}^{\ominus}$。

五、注意事项

1. 本实验所用去离子水的电导率小于 $5\times 10^{-4} S/m$ 时,可使 $BaSO_4$ 的溶度积接近文献值($1\mu S/cm=1.0\times 10^{-4} S/m$)。

2. 注意水的纯度不高或所用玻璃器皿不够洁净,都将对实验结果产生影响。

六、思考题

1. 制备 $BaSO_4$ 时,为什么要多次洗涤至无 Cl^-?
2. 为什么纯水具有一定的导电能力?
3. 什么情况下电解质的摩尔电导率是其离子摩尔电导率的简单加和?

Experiment 5 Determination of the Solubility Product Constant of BaSO₄ (Electrical Conductivity Method)

1. Purpose

(1) Learn the basic operations of precipitation generation, aging, centrifugation, washing, etc.

(2) Master the method for measuring the solubility product of insoluble electrolytes.

(3) Learn how to use a conductivity meter.

2. Principle

In a saturated solution of insoluble electrolyte $BaSO_4$, there exists the following equilibrium:

$$BaSO_4 \rightleftharpoons Ba^{2+} + SO_4^{2-}$$

Its solubility product is: $K_{sp(BaSO_4)}^{\ominus} = \left\{\dfrac{[Ba^{2+}]}{c^{\ominus}}\right\}\left\{\dfrac{[SO_4^{2-}]}{c^{\ominus}}\right\}$

Due to the low solubility of insoluble electrolyte, it is difficult to directly measure its K_{sp}. But as long as there is dissolution, there are ionized charged ions in the solution. Therefore, the solubility of the insoluble electrolyte can be calculated by measuring the conductance or conductivity of the solution, and then converting it to a solubility product based on its relationship with concentration.

The conductivity of an electrolyte solution can be expressed as resistance R or conductivity G (S), which are reciprocal to each other.

$$G = \frac{1}{R}$$

Because $R = \rho \frac{l}{A}$

So
$$G = \frac{1}{\rho} \cdot \frac{A}{l} = \kappa \cdot \frac{A}{l}$$

Where ρ is resistivity; l is resistance length; A is cross sectional area of resistance; κ is conductivity, in units of S/m.

Conductivity represents the conductivity of a solution placed between two electrodes with an area of $1m^2$ and a distance of $1m$. Molar conductivity (Λ_m) is the conductivity of a solution containing 1 mol of electrolyte placed between electrodes at a distance of 1 m.

$$\Lambda_m = \frac{\kappa}{c} \; [(S \cdot m^2)/mol]$$

Ultimate molar conductivity Λ_∞ represents the molar conductivity under infinite dilution conditions. Due to the extremely low solubility of $BaSO_4$, its solution can be regarded as an infinitely diluted solution. So Λ_∞ is a constant.

$$\Lambda_{mBaSO_4} = \Lambda_{mBa^{2+}} + \Lambda_{mSO_4^{2-}}$$

$$\Lambda_{\infty BaSO_4} = \Lambda_{mBaSO_4} \quad c_{BaSO_4} = \frac{\kappa_{BaSO_4}}{1000\Lambda_{\infty BaSO_4}}$$

Because:
$$c_{BaSO_4} = c_{Ba^{2+}} = c_{SO_4^{2-}}$$

$$K^\ominus_{sp(BaSO_4)} = [Ba^{2+}][SO_4^{2-}] = c^2_{BaSO_4} = \left(\frac{\kappa_{BaSO_4}}{1000\Lambda_{\infty BaSO_4}}\right)^2 = \left[\frac{\kappa_{BaSO_4(Solution)} - \kappa_{H_2O}}{1000\Lambda_{\infty BaSO_4}}\right]^2$$

$$\Lambda_{\infty BaSO_4} = 2.872 \times 10^{-2} (S \cdot m^2)/mol$$

3. Apparatus and Chemicals

Apparatus: Conductivity meter; centrifuge; alcohol lamp; 100mL beaker; surface plate; glass rod.

Chemicals: H_2SO_4 (0.05mol/L); $BaCl_2$ (0.05mol/L); $AgNO_3$ (0.01mol/L).

4. Procedure

(1) Preparation of $BaSO_4$ precipitate

Take 30mL of 0.05mol/L H_2SO_4 from a measuring cylinder and place it in a 100mL small beaker. Heat the H_2SO_4 solution with an alcohol lamp and bring it to a boil. Stir while adding 30mL of 0.05mol/L $BaCl_2$ solution. After adding the solution, wash the beaker containing $BaCl_2$ solution with 5mL of deionized water and pour it all into the mixture. Cover it with the watch glass. Continue heating and keep boil for 5minutes, keep warm on low heat for 10minutes, remove and let stand for 15-20minutes, then pour out the upper layer of clear liquid. Stir the precipitate and a small amount of residual liquid into an emulsion using a glass rod, transfer them in batches to a centrifuge tube, perform centrifugation separation (3000r/min, 5min), and discard the upper solution.

(2) Configuration of $BaSO_4$ saturated solution

Add 40mL of deionized water to a small beaker and heat it to near boiling for washing the $BaSO_4$ precipitate. When washing, add 5mL of hot deionized water each time, stir the precipitate thoroughly with a glass rod, centrifuge and separate, and discard the washing solution. After washing three times, pour the washing solution into a clean beaker and verify the absence of Cl^- in the washing solution with 0.01mol/L $AgNO_3$ solution (if precipitation occurs, continue washing until no precipitation is generated by adding $AgNO_3$ solution dropwise to the washing solution). Add a small amount of deionized water to the $BaSO_4$ precipitate, stir well with a glass rod, and transfer to a small beaker of 50mL of deionized water with measured conductivity. Cover with a watch glass, heat to boil for 3-5minutes, and continuously stir for 5minutes. Let stand, cool, and transfer the supernatant to the beaker for measuring conductivity.

(3) Determination of conductivity

Measure the conductivity of deionized water and $BaSO_4$ saturated solution separately (κ, $\mu S/cm$), calculate $K_{sp}^{\ominus}$ of $BaSO_4$.

5. Notes

(1) If the conductivity of the distilled water in this experiment is less than $5 \times 10^{-4} S/m$, the solubility product of $BaSO_4$ will be approached to the literature value ($1\mu S/cm = 1.0 \times 10^{-4} S/m$).

(2) Note that low purity of water or unclean glassware used will have an impact on the experimental results.

6. Questions

(1) Why should Cl^- be washed out in the preparation of $BaSO_4$?

(2) Why does pure water have a certain degree of conductivity?

(3) Under what circumstances is the molar conductivity of an electrolyte a simple sum of its ionic molar conductivity?

实验六 碱金属和碱土金属元素鉴定

一、实验目的

1. 了解金属钠和镁的强还原性。
2. 学会用焰色反应鉴定某些碱金属和碱土金属离子的方法。
3. 掌握碱土金属难溶盐的溶解性。

二、仪器与药品

仪器：离心机；小试管；小刀；镊子；研钵；坩埚；铂丝或镍铬丝；pH试纸；钴玻璃等。

药品：HCl（2mol/L，6mol/L）；HNO_3（6mol/L）；H_2SO_4（2mol/L）；HAc（2mol/L）；NaOH（2mol/L）；Na_2CO_3（0.1mol/L）；$NH_3 \cdot H_2O\text{-}NH_4Cl$ 缓冲溶液（浓度各为1mol/L）；$HAc\text{-}NH_4Ac$ 缓冲溶液（浓度各为1mol/L）；$MgCl_2$（0.1mol/L）；$CaCl_2$（0.1mol/L）；$BaCl_2$（0.1mol/L）；Na_2SO_4（0.5mol/L）；$CaSO_4$（饱和）；$(NH_4)_2C_2O_4$（饱和）；$KMnO_4$（0.01mol/L）；$(NH_4)_2CO_3$（0.5mol/L）；K_2CrO_4（0.1mol/L）；Na^+、K^+、Ca^{2+}、Sr^{2+}、Ba^{2+} 试液（10g/L）；酚酞溶液；Na(s)；Mg(s)；镁试剂Ⅰ等。

三、实验步骤

1. 金属钠和镁的还原性

（1）金属钠和氧的反应

用镊子夹取一小块金属钠，用滤纸吸干其表面的煤油，放入干燥的坩埚中加热。当钠开始燃烧时，停止加热，观察反应现象及产物的颜色和状态。

（2）镁条在空气中燃烧

取一小段镁条，用砂纸除去表面的氧化物，点燃，观察燃烧情况和所得产物。产物中可能存在 Mg_3N_2 吗？如何证实？

（3）钠、镁与水的反应

取一小块金属钠，用滤纸吸干其表面煤油，放入盛有1/4体积水的250mL烧杯中，观察反应情况。检验反应后水溶液的酸碱性。另取一段擦至光亮的镁条，投入盛有2mL去离子水的试管中，观察反应情况。水浴加热，反应是否明显？检验反应后水溶液的酸碱性。

2. 碱金属和碱土金属的焰色反应

碱金属和碱土金属的挥发性化合物在氧化焰上灼烧时，能使火焰呈现特殊的颜色。例如，钠呈现黄色、钾呈现紫色、钙呈现橙色、锶呈现深红色、钡呈现黄绿色。因此，分析化学中常借此鉴定这些元素，并称之为焰色反应。

实验：取镶有铂丝（或镍铬丝）的玻璃棒一根（金属丝的尖端弯成环状），先按下法清洁之：浸铂丝于纯的6mol/L HCl中（放在点滴板的凹穴内），在煤气灯的氧化焰上灼烧片刻，再浸入酸中，取出再灼烧，如此重复数次，直至火焰无色（用镍铬丝时，仅能烧至呈淡黄色）。这时铂丝才算洁净。

用洁净的铂丝蘸取 Na^+ 试液（预先放在点滴板的凹穴内加1滴6mol/L HCl）灼烧之，观察火焰的颜色。

用与上面相同的操作，分别观察钾、钙、锶和钡等盐溶液的焰色反应（观察钾盐的焰色反应时，为消除钠对钾焰色的干扰，一般需用蓝色钴玻璃片滤光）。

3. 镁、钙和钡的难溶盐的生成和性质

（1）硫酸盐溶解度的比较

在三支试管中，分别加入1mL 0.1mol/L $MgCl_2$、$CaCl_2$ 和 $BaCl_2$ 溶液，再加入1mL 0.5mol/L 的 Na_2SO_4 溶液，有何现象（若无沉淀生成，稍微加热后再观察）？分离出沉淀，试验其与 HNO_3（6mol/L）的作用。

另取两支试管，分别加入1mL 0.1mol/L $MgCl_2$ 和 $BaCl_2$ 溶液，然后各加0.5mL饱和硫酸钙溶液，又有何现象？比较 $MgSO_4$、$CaSO_4$ 和 $BaSO_4$ 的溶解度。

（2）镁、钙和钡的碳酸盐的生成和性质

a. 在三支试管中，分别加入0.5mL 0.1mol/L 的 $MgCl_2$、$CaCl_2$ 和 $BaCl_2$ 溶液，再各

入 0.5mL 0.1mol/L 的 Na_2CO_3 溶液，稍加热，观察现象。试验产物对 2mol/L NH_4Cl 溶液的作用，写出反应式。

b. 在三支试管中，分别加入 0.5mL 0.1mol/L 的 $MgCl_2$、$CaCl_2$ 和 $BaCl_2$ 溶液，再各加入 0.5mL $NH_3 \cdot H_2O$-NH_4Cl 缓冲溶液（pH = 9），然后各加入 0.5mL 0.5mol/L $(NH_4)_2CO_3$ 溶液。稍加热，观察现象。试指出 Mg^{2+} 与 Ca^{2+}、Ba^{2+} 的分离条件。

(3) 钙和钡的铬酸盐的生成和性质

a. 在两支试管中，各加入 0.5mL 0.1mol/L 的 $CaCl_2$ 和 $BaCl_2$ 溶液，再各加入 0.5mL 0.1mol/L K_2CrO_4 溶液，观察现象。试验产物对 2mol/L HAc 溶液的作用，写出反应式。

b. 在两支试管中，各加入 0.5mL 0.1mol/L 的 $CaCl_2$ 和 $BaCl_2$ 溶液，再各加入 0.5mL HAc-NH_4Ac 缓冲溶液，然后各加入 0.5mL 0.1mol/L K_2CrO_4 溶液，观察现象。试指出 Ca^{2+} 和 Ba^{2+} 的分离条件。

四、注意事项

金属钠和钙应保存在煤油或石蜡油中。取用时，可在煤油中用小刀切割，用镊子夹取，并用滤纸把煤油吸干。切勿与皮肤接触，未用完的钠屑不能乱丢，可放回原瓶中或放在少量酒精中，使其缓慢耗掉。

五、思考题

1. 如何从碳酸钙和碳酸钡沉淀中分离钙离子和钡离子？
2. 若实验室中发生镁燃烧事故，应用什么方法灭火？可否用水或二氧化碳来灭火？

Experiment 6　Identification of Alkali Metal and Alkaline Earth Metal Elements

1. Purpose

(1) To understand the reactivity of sodium and magnesium.

(2) To master the operations of flame reactions.

(3) To test and compare the insolubility of hydroxides of the alkali earth metals.

2. Apparatus and Chemicals

Apparatus: Centrifuge; small test tubes; knife; forceps; mortar box; melting pot; platinum filament (or nickel-chromium filament); pH test paper; blue cobalt glass and so on.

Chemicals: HCl (2mol/L, 6mol/L); HNO_3 (6mol/L); H_2SO_4 (2mol/L); HAc (2mol/L); NaOH (2mol/L); Na_2CO_3 (0.1mol/L); $NH_3 \cdot H_2O$-NH_4Cl buffer solution (1mol/L); HAc-NH_4Ac buffer solution (1mol/L); $MgCl_2$ (0.1mol/L); $CaCl_2$ (0.1mol/L); $BaCl_2$ (0.1mol/L); Na_2SO_4 (0.5mol/L); $CaSO_4$ (saturated solution); $(NH_4)_2C_2O_4$ (saturated solution); $KMnO_4$ (0.01mol/L); $(NH_4)_2CO_3$ (0.5mol/L); K_2CrO_4 (0.1mol/L); Na^+, K^+, Ca^{2+}, Sr^{2+}, Ba^{2+} solution (10g/L); phenolphthalein;

Na(s); Mg(s); azoviolet Ⅰ and so on.

3. Procedure

3.1 Reducibility of sodium and magnesium

(1) Reaction of sodium with oxygen in air

A piece of sodium of soybean size is cut with scissors; the kerosene on its surface is absorbed by filer paper, then immediately put it into a crucible, heat; when the sodium begins burning, stop heating. Observe the color and state of products.

(2) Reaction of magnesium with oxygen in air

Take magnesium of 1 cm length, remove the superficial membrane with sand paper, light it and observe the phenomena of the reaction. Is there compound of Mg_3N_2 in products? How to test it?

(3) Reaction of sodium and magnesium with water

A piece of sodium of soybean size is cut with scissors, the kerosene on its surface is absorbed by filter paper, then put it into a 250mL beaker with 1/4 volume of water. Observe the phenomenon of the reaction and test the acidity or the alkalinity of the solution. Put polished magnesium ribbon into a test tube with 2mL distilled water. Observe the phenomenon of the reaction. Heat the test tube, what will happen? Test the acidity or the alkalinity of the solution.

3.2 Flame test of alkali metals, alkali earth metals

A flame test is an analytic procedure used in chemistry to detect the presence of certain elements, primarily alkali metal and alkali earth metal ions, based on each element's characteristic emission spectrum. The color of flames in general also depends on temperature; see flame color, for example: Na-yellow; K-lilac; Ca-brick red; Sr-crimson; Ba-apple green.

Take a glass rod with a platinum wire (or nickel-chromium wire). To clean the wire whose tip is bent into a ring, dip it into the test tube of 6mol/L HCl and heat the wire on the oxidizing flame. Repeat this operation until the flame appears colorless (or pale yellow for nickel-chromium wire). When the platinum wire is clean, dip the wire in the test tube containing Na^+, K^+, Ca^{2+}, Sr^{2+}, Ba^{2+} solution respectively. Every time the wire must be cleaned according to the method above. As we observe the color of the flame of the potassium ion, a piece of blue cobalt glass should be used.

3.3 The properties, formation of the insoluble salts of calcium, magnesium, barium

(1) Compare the solubility of sulfate salts of calcium, magnesium, barium

Add 1mL of 0.1mol/L $MgCl_2$, $CaCl_2$ and $BaCl_2$ solution to three test tubes respectively, then add 1mL 0.5mol/L Na_2SO_4 solution respectively. Observe the phenomena. If there is no precipitate, heat it gently. Test the reaction of the precipitate with 6mol/L HNO_3.

In other two tubes, 1mL of 0.1mol/L $MgCl_2$ and $BaCl_2$ are added respectively first, then 0.5mL $CaSO_4$ (saturated solution) are added, observe the formation of precipitate. Compare the solubility of $MgSO_4$, $BaSO_4$ and $CaSO_4$.

(2) The properties, formation of calcium carbonate, barium carbonate and magnesium carbonate

a. Add 0.5mL of 0.1mol/L $MgCl_2$, $CaCl_2$ and $BaCl_2$ solution to three different tubes, then add 0.5mL 0.1mol/L Na_2CO_3 solution to each tube and heat them for a little while. Observe the phenomena. Test the reaction between the product and 2mol/L NH_4Cl and write down the reaction equations.

b. Add 0.5mL of 0.1mol/L $MgCl_2$, $CaCl_2$ and $BaCl_2$ solution to three different tubes, then add 0.5mL $NH_3 \cdot H_2O$-NH_4Cl buffer solution (pH=9) and 0.5mL 0.5mol/L $(NH_4)_2CO_3$ solution to each tube. Heat them for a little while and observe the phenomena. Summarize the method and the condition of the separation of Mg^{2+}, Ca^{2+} and Ba^{2+}.

(3) The properties, formation of calcium chromate, barium chromate

a. Add 0.5mL of 0.1mol/L $CaCl_2$ and $BaCl_2$ solution to two different tubes, then add 0.5mL 0.1mol/L K_2CrO_4 solution respectively. Observe the phenomena. Test the reaction of the product with 2mol/L HAc solution and write down the reaction equations.

b. Add 0.5mL of 0.1mol/L $CaCl_2$ and $BaCl_2$ solution to two different tubes, then add 0.5mL HAc-NH_4Ac buffer solution and 0.5mL 0.1mol/L K_2CrO_4 solution respectively. Observe the phenomena and summarize the condition of separating Ca^{2+} and Ba^{2+}.

4. Notes

Sodium and calcium should be stored in kerosene or liquid paraffin. As we use them, we should cut them into small ones in the kerosene. Fetch it with forceps, and absorb the kerosene with filter paper. Make sure not to contact with skin. Residual sodium should not be thrown over casually, you may add a small amount of anhydrous ethanol to make it decompose slowly or recycle.

5. Questions

(1) How do we separate Ca^{2+}, Ba^{2+} from $CaCO_3$, $BaCO_3$ precipitate?

(2) If there is a fire accident caused by magnesium in lab, how shall we put out the fire? Can we use the water or carbon dioxide to put it out?

实验七　硼族元素、碳族元素和氮族元素鉴定

一、实验目的

1. 掌握硼酸和硼砂的主要性质。
2. 试验并掌握锡（Ⅱ）的还原性和铅（Ⅳ）的氧化性。
3. 掌握铵盐、亚硝酸、硝酸和磷酸盐的主要性质。

二、仪器与药品

仪器：离心机；小试管；pH试纸；水浴锅等。

药品：$SnCl_2$（0.1mol/L）；NaOH（2mol/L）；$Bi(NO_3)_3$（0.1mol/L）；HNO_3（6mol/L，2mol/L）；PbO_2(s)；硼砂饱和溶液；浓 H_2SO_4；H_3BO_3 固体；甘油；甲基橙指示剂；NH_4Cl 固体；$NH_4H_2PO_4$ 固体；NH_4NO_3 固体；饱和 $NaNO_2$ 溶液；KI（0.1mol/L）；$NaNO_2$（0.5mol/L）；$KMnO_4$（0.01mol/L）；浓 HNO_3；硫粉；铜片；石蕊试纸；Na_2HPO_4（0.1mol/L）；钼酸铵试剂；$Na_4P_2O_7$（0.1mol/L）；$NaPO_3$（0.1mol/L）；$AgNO_3$（0.1mol/L）；HAc（2mol/L）；蛋白溶液。

三、实验步骤

1. Sn（Ⅱ），Pb（Ⅳ）的氧化还原性质

a. 亚锡酸钠的还原性。在盛有 1mL 0.1mol/L $SnCl_2$ 溶液的试管中，滴加 2mol/L NaOH 溶液，同时不断振荡，直至生成的沉淀完全溶解再过量 3 滴，然后加入 0.1mol/L $Bi(NO_3)_3$ 溶液数滴，有何现象？写出反应方程式，此反应可用于鉴定 Bi^{3+}。

b. PbO_2 的氧化性。取 1 滴 0.1mol/L $MnSO_4$ 溶液于试管中，加水 10 滴稀释，再加 1mL 6mol/L HNO_3 和 PbO_2 固体少许，搅拌后置水浴中加热，有何变化？写出反应式。

2. 硼酸的制备和性质

a. 硼酸的生成。取 1mL 硼砂饱和溶液，测其 pH。在该溶液中加入 0.5mL 浓 H_2SO_4，用冰水冷却之，有无晶体析出？离心分离，弃去溶液。用少量冷水洗涤晶体 2~3 次，再用 0.5mL 水使之溶解，用 pH 试纸测其 pH，与硼砂溶液相比是否相同？

b. 硼酸的性质。试管中加入少量 H_3BO_3 固体和 6mL 去离子水，微热之，使固体溶解。加一滴甲基橙指示剂，观察溶液的颜色。

把溶液分装于两支试管中，在一支试管中加几滴甘油 $C_3H_5(OH)_3$，混匀，比较两支试管的颜色，解释之。

硼酸和甘油的反应为

$$HO-B\begin{matrix}OH\\OH\end{matrix} + \begin{matrix}HO-CH_2\\CHOH\\HO-CH_2\end{matrix} \longrightarrow \left[O-B\begin{matrix}O-CH_2\\\\O-CH_2\end{matrix}CHOH\right]^- + H^+ + 2H_2O$$

3. 铵盐的热分解与阴离子的关系

a. 阴离子为挥发性酸根。在干燥试管内放入约 1 g 的 NH_4Cl 固体，加热试管底部（底部略高于管口），用潮湿的红色石蕊试纸在管口检验逸出气体，观察试纸颜色的变化。继续加强热，石蕊试纸又怎样变化？观察试管上部冷壁上有白霜出现，解释实验过程中所出现的现象。

b. 阴离子为不挥发性酸根。在干燥试管中加热 $NH_4H_2PO_4$ 固体，检验释放的气体为何物？

c. 阴离子为氧化性酸根。取少量 NH_4NO_3 固体放在干燥试管内，加热观察现象。

总结铵盐的热分解产物与阴离子的关系，写出 NH_4Cl、$NH_4H_2PO_4$ 和 NH_4NO_3 的热分解反应方程式。

4. 亚硝酸的生成和性质

（1）亚硝酸的生成

将在冰水中冷却过的 1mL 饱和 $NaNO_2$ 溶液和 1mL 1mol/L H_2SO_4 于试管中混合，有

何现象？溶液放置一段时间，有何现象发生？写出反应方程式，解释之。

(2) NO_2^- 的氧化性和还原性

a. 在盛有 0.5mL 0.1mol/L KI 溶液的试管中，加入几滴 1mol/L H_2SO_4 酸化，再加入几滴 0.5mol/L $NaNO_2$ 溶液，摇动，观察溶液颜色的变化和气体的放出，检验之。

b. 在盛有 0.5mL 0.01mol/L $KMnO_4$ 溶液的试管中，加几滴 1mol/L H_2SO_4 酸化，再加入几滴 0.5mol/L $NaNO_2$ 溶液，振荡，有何现象？

查出有关电对的标准电极电势，写出酸化的 $KMnO_4$ 溶液和 KI 溶液与 $NaNO_2$ 溶液的反应方程式，指出 $NaNO_2$ 是氧化剂还是还原剂。

5. 硝酸的氧化性

a. 浓 HNO_3 与非金属的作用。在小试管内放入少许硫粉，加入浓 HNO_3 10 滴，水浴加热。待硫大部分溶解后用滴管取出溶液少许放在另一小试管中，用少量去离子水稀释后加几滴 0.1mol/L $BaCl_2$ 溶液，有何现象？硫的氧化产物是什么？写出反应式。

b. 浓 HNO_3 与金属的作用。取一小块铜片放入小试管中，滴加 0.5mL 浓 HNO_3，注意观察放出气体的颜色。写出反应式。

c. 稀 HNO_3 与金属的作用。取一小块铜片放入小试管中，滴加 0.5mL 6mol/L HNO_3，水浴上微热，注意观察与上一反应现象有何异同？在试管口气体的颜色有无变化？写出反应式。

d. 稀硝酸与活泼金属的作用。取一小段镁条放入小试管中，加入 1mL 1mol/L HNO_3，有何现象？用检验 NH_4^+ 的方法检验溶液中是否有 NH_4^+ 生成（检验方法参看附录四）。

6. 各种磷的含氧酸根的区别与鉴定

a. 磷的含氧酸根的鉴定——磷钼酸铵沉淀法。PO_4^{3-} 与钼酸铵试剂（钼酸铵在硝酸中的溶液）生成特殊的黄色晶状磷钼酸铵 $(NH_4)_3P(Mo_3O_{10})_4$ 沉淀：

$$PO_4^{3-} + 3NH_4^+ + 12MoO_4^{2-} + 24H^+ \Longrightarrow (NH_4)_3P(Mo_3O_{10})_4 + 12H_2O$$

此反应可用于检验 PO_3^-，$P_2O_7^{4-}$ 或 PO_4^{3-}。若在冷溶液中生成黄色沉淀，可判断 $H_2PO_4^-$，HPO_4^{2-} 或 PO_4^{3-} 的存在；若在冷溶液中无沉淀生成，经加热后可得黄色沉淀，可判断 PO_3^- 或 $P_2O_7^{4-}$ 的存在，因为加热可使它们转化为 PO_4^{3-}。

取 0.1mol/L Na_2HPO_4 溶液 2 滴，加入 8~10 滴钼酸铵试剂，用玻璃棒摩擦管壁，有黄色磷钼酸铵生成，表示有 PO_4^{3-} 存在。另取 0.1mol/L $Na_4P_2O_7$ 或 $NaPO_3$ 溶液进行同上实验，若无黄色沉淀产生，可在水浴上微热片刻，有无变化？说明变化原因。

b. 磷的含氧酸根与 $AgNO_3$ 溶液的作用。取 0.1mol/L 的 Na_2HPO_4，$Na_4P_2O_7$ 与 $NaPO_3$ 溶液各 2 滴分装于 3 支试管中，向各管加入 0.1mol/L $AgNO_3$ 溶液 2~3 滴，有何现象产生？再在各管中加入少量 2mol/L HNO_3，沉淀有无变化？

c. 磷的含氧酸根与蛋白溶液的作用。取 0.1mol/L 的 Na_2HPO_4，$Na_4P_2O_7$ 与 $NaPO_3$ 溶液各 2 滴分装于 3 支试管中，加少许 2mol/L HAc 溶液，使溶液呈酸性，各加入蛋白溶液 10 滴，振荡。观察各试管中蛋白溶液是否有凝固现象？

四、思考题

1. 如何分别检出 $NaNO_2$、$Na_2S_2O_3$ 和 KI 溶液？
2. 设计三种区别亚硝酸钠和硝酸钠的方案。

3. 如何消除 NO_2^- 对鉴定 NO_3^- 的影响？

Experiment 7　Identification of Boron, Carbon, and Nitrogen Group Elements

1. Purpose

(1) To master the main properties of boric acid and borax.

(2) To test and master the reducibility of tin (Ⅱ) and oxidizability of lead (Ⅳ).

(3) To master the main properties of ammonium salt, nitric acid, nitrous acid and phosphate.

2. Apparatus and Chemicals

Apparatus: Centrifuge; small test tubes; pH test paper; water bath and so on.

Chemicals: $SnCl_2$ (0.1mol/L); NaOH (2mol/L); $Bi(NO_3)_3$ (0.1mol/L); HNO_3 (6mol/L, 2mol/L); PbO_2 (s); saturated borax solution; concentrated H_2SO_4; H_3BO_3 (s); glycerol $C_3H_5(OH)_3$; methyl orange; NH_4Cl(s); $NH_4H_2PO_4$(s); NH_4NO_3(s); saturated $NaNO_2$ solution; KI (0.1mol/L); $NaNO_2$ (0.5mol/L); $KMnO_4$ (0.01mol/L); concentrated HNO_3; sulfur powder; copper sheet; red litmus test paper; Na_2HPO_4 (0.1mol/L); ammonium molybdate; $Na_4P_2O_7$ (0.1mol/L); $NaPO_3$ (0.1mol/L); $AgNO_3$ (0.1mol/L); HAc (2mol/L); protein solution.

3. Procedure

3.1　Reducibility of tin (Ⅱ) and oxidizability of lead (Ⅳ)

a. Reducibility of tin (Ⅱ). Drop 2mol/L NaOH solution to a test tube with 1mL 0.1mol/L $SnCl_2$ solution and shake it until the product is dissolved. Three more drops of NaOH are added. Then place drops of 0.1mol/L $Bi(NO_3)_3$ solution. Observe the phenomena and write down the reaction equations. (This reaction can be used to identify Bi^{3+}.)

b. The oxidizability of PbO_2. Add one drop of 0.1mol/L $MnSO_4$ and 10 drops of distilled water into the test tube. Then add 1mL 6mol/L HNO_3 solution and a small amount of solid PbO_2. What will happen when heat and stir it under water bath. Then write down the reaction equations.

3.2　Preparation, properties of boric acid

a. Preparation of boric acid. Take 1mL saturated borax solution, then test its pH value by pH test paper. Add 0.5mL concentrated H_2SO_4, cool it down in ice water. Observe whether there are crystals. Centrifugalize the tube, remove solution of above layer, wash the precipitate with a little of ice water for 2-3 times. Then dissolve the precipitate with 0.5mL distilled water and test pH value again. Compare the mixed solution with the borax solution.

b. Properties of boric acid. Add a small amount of solid H_3BO_3 and 6mL distilled water to the test tube. Heat gently until the solid is dissolved. Add a drop of methyl orange solution and observe the color.

Divide the solution into two different test tubes. Add drops of glycerol $C_3H_5(OH)_3$ to one of them, shake it and compare the color of the solution of the two tubes. Then explain the reason in the report.

The reaction mechanism of boric acid and glycerol is:

$$HO-B(OH)_2 + \begin{matrix} HO-CH_2 \\ CHOH \\ HO-CH_2 \end{matrix} \longrightarrow \left[O-B \begin{matrix} O-CH_2 \\ O-CH_2 \end{matrix} CHOH \right]^- + H^+ + 2H_2O$$

3.3 Thermal decomposition of ammonium salts

a. The anions are volatile acid radicals. Place about 1g solid NH_4Cl in a dry test tube, and put a red litmus test paper at the opening of the test tube. Heat the bottom of the test tube slight. Observe the color change of the test paper. What is white compound in the test tube after being cooled down? Record and explain in your report.

b. The anions are non-volatile acid radicals. Use $NH_4H_2PO_4$ to replicate the above experiment and test the produced gas.

c. The anions are oxidizing acid radicals. Use NH_4NO_3 to replicate the above experiment and observe the phenomenon.

Write down the reaction equations, and sum up the relationship between the decomposition product of ammonium salts and the anions.

3.4 Properties, formation of nitrite

(1) The formation of nitrite. Mix 1mL saturated $NaNO_2$ solution which has been cooled down by ice-water and about 1mL 1mol/L H_2SO_4. Observe the phenomenon. What will happen when the solution is leaved alone for a while? Write down the reaction equations and explain in your report.

(2) Oxidizability and reducibility of NO_2^-

a. Acidify 0.5mL 0.1mol/L KI solution with drops of 1mol/L H_2SO_4, and then add several drops of 0.5mol/L $NaNO_2$ solution. Shake it, observe the color change and test the produced gas.

b. Acidify 0.5mL 0.01mol/L $KMnO_4$ solution with drops of 1mol/L H_2SO_4, and then add several drops of 0.5mol/L $NaNO_2$ solution. Shake it and observe what change should take place. Write down the reaction equations. According to the standard electrode potentials, point out the $NaNO_2$ is oxidizing agent or reducing regent in the reactions above.

3.5 Oxidizability of nitric acid

a. Reactions of concentrated nitric acid with nonmetals. Put a small amount of sulfur powder into a test tube, then add 10 drops of concentrated HNO_3 and heat it under water bath until the solid is dissolved. Take a small amount of solution with a long burette and put it in another test tube. Add drops of 0.1mol/L $BaCl_2$ solution after diluting it with distilled water. What will happen and what is the oxidation product of sulfur? Write down the reaction equations.

b. Reactions of concentrated nitric acid with metals. Put a small piece of copper sheet into the test tube. Add 0.5mL concentrated HNO_3. Observe the color of the produced gas and write down the reaction equations.

c. Reactions of dilute nitric acid with metals. Take a small piece of copper sheet into the test tube. Add 0.5mL 6mol/L HNO_3 and heat it under the water bath. Observe the phenomena and tell the difference between the previous reactions in a. and b. Is there any color change of the produced gas at the opening of the test tube? Write down the reaction equations.

d. Reactions of dilute nitric acid with active metals. Take a piece of magnesium and 1mL 1mol/L HNO_3 solution in the test tube. What will happen? Test whether the NH_4^+ exist in the solution according to the identification method listed in appendix 4. Write down the reaction equations.

3.6 Identification, difference of oxyacid radicals of phosphorus

a. Identification of oxyacid radicals of phosphorus—ammonium molybdate method. PO_3^-, $P_2O_7^{4-}$ and PO_4^{3-} can all react with ammonium molybdate $(NH_4)_2MoO_4$ to produce yellow precipitates $(NH_4)_3P(Mo_3O_{10})_4$.

$$PO_4^{3-} + 3NH_4^+ + 12MoO_4^{2-} + 24H^+ = (NH_4)_3P(Mo_3O_{10})_4 \downarrow + 12H_2O$$

The yellow precipitates in cooled solution confirm that $H_2PO_4^-$, HPO_4^{2-} or PO_4^{3-} are present. The appearance of yellow precipitates after heating the cooled solution confirms that PO_3^- or $P_2O_7^{4-}$ are present.

Mix two drops of 0.1mol/L Na_2HPO_4 solution and 8-10 drops of ammonium molybdate $(NH_4)_2MoO_4$ in the test tube. Rub the inside of the test tube with a glass rod until there are yellow precipitates, which can prove the existence of PO_4^{3-}. Take 0.1mol/L $Na_4P_2O_7$ or $NaPO_3$ in the test tube and repeat the experiment above. Heat it gently under water bath for a while if there is not yellow precipitate. What will happen? Explain the phenomena in your report.

b. Reactions of oxyacid radicals of phosphorus with $AgNO_3$. Place two drops of 0.1mol/L Na_2HPO_4, $Na_4P_2O_7$ and $NaPO_3$ to three different test tubes, and add 2-3 drops of 0.1mol/L $AgNO_3$ solution respectively. What is the phenomenon? Observe the color and state of the precipitate after adding a small amount of 2mol/L HNO_3 to the three test tube respectively.

c. Reactions of oxyacid radicals of phosphorus with protein solution. Take two drops of 0.1mol/L Na_2HPO_4, $Na_4P_2O_7$ and $NaPO_3$ to three test tubes, and acidify them with 2mol/L HAc respectively. Add 10 drops of protein solution to the test tubes and shake them. Observe whether there is the coagulation of protein solution in the tubes.

4. Questions

(1) How to identify $NaNO_2$, $Na_2S_2O_3$ and KI solution?

(2) Design three methods to identify NaNO$_2$ and NaNO$_3$.

(3) When identifying NO$_3^-$, how do we remove NO$_2^-$?

实验八　氧族元素和卤族元素鉴定

一、实验目的

1. 了解过氧化氢和硫代硫酸钠的性质，掌握过氧化氢的鉴定方法。
2. 掌握 SO_4^{2-}、SO_3^{2-}、$S_2O_3^{2-}$、S^{2-} 的鉴定方法。
3. 掌握卤素含氧酸盐的氧化性。
4. 了解某些金属卤化物的性质。

二、仪器与药品

仪器：小试管；水浴锅等。

药品：H$_2$SO$_4$（2mol/L，6mol/L，3mol/L）；HCl（2mol/L，6mol/L）；K$_2$CrO$_4$（0.1mol/L）；NaOH（2mol/L）；Pb(NO$_3$)$_2$（0.1mol/L）；BaCl$_2$（0.1mol/L）；H$_2$O$_2$（w 为 0.03）；AgNO$_3$（0.1mol/L）；K$_4$[Fe(CN)$_6$]溶液（0.1mol/L）；Na$_2$S$_2$O$_3$（0.1mol/L，0.5mol/L）；ZnSO$_4$（饱和）；MnSO$_4$（0.1mol/L）；Na$_2$[Fe(CN)$_5$NO]溶液；氯水；碘水；硫代乙酰胺（w 为 0.05）；MnO$_2$(s)；乙醚；KClO$_3$ 晶体；浓 HCl；饱和 KClO$_3$ 溶液；Na$_2$SO$_3$（0.1mol/L）；四氯化碳；KI（0.1mol/L，0.5mol/L）；饱和 KBrO$_3$ 溶液；KBr（0.1mol/L，0.5mol/L）；NaF（0.1mol/L）；NaCl（0.1mol/L）；KI（0.1mol/L）；Ca(NO$_3$)$_2$（0.1mol/L）；HNO$_3$（2mol/L）；NH$_3$·H$_2$O（2mol/L）；淀粉溶液等。

三、实验步骤

1. 过氧化氢的鉴定

取 w 为 0.03 的 H$_2$O$_2$ 溶液 2mL 于一试管中，加入 0.5mL 乙醚和 1mL 2mol/L H$_2$SO$_4$，再加入 3~5 滴 0.1mol/L K$_2$CrO$_4$ 溶液，观察水层和乙醚层中的颜色变化。根据实验证明上述实验制得的是过氧化氢溶液。

2. 过氧化氢的性质

a. 过氧化氢的氧化性。在小试管中加入几滴 0.1mol/L Pb(NO$_3$)$_2$ 溶液和 w 为 0.05 的硫代乙酰胺溶液，在水浴上加热，有何现象？离心分离，弃去溶液，并用少量去离子水洗涤沉淀 2~3 次，然后往沉淀中加入少许 w 为 0.03 的过氧化氢溶液，沉淀有何变化？解释之。

b. 过氧化氢的还原性。在试管里加入 0.5mL 0.1mol/L AgNO$_3$ 溶液，然后滴加 2mol/L NaOH 溶液至有沉淀产生。再往试管中加入少量 w 为 0.03 的过氧化氢溶液，有何现象？注意产物颜色有无变化并用带余烬的火柴检验，有何种气体产生？试解释之。

c. 介质酸碱性对过氧化氢氧化还原性质的影响。取 1mL w 为 0.03 的 H$_2$O$_2$ 溶液于试管中，加入 2mol/L NaOH 溶液数滴，再加入 0.1mol/L MnSO$_4$ 溶液数滴，充分振荡后观察现象。溶液静置后除去上层清液，往沉淀中加入少量 2mol/L H$_2$SO$_4$ 溶液后，滴加 w 为 0.03 的 H$_2$O$_2$ 溶液，观察又有什么现象发生？写出反应方程。

d. 过氧化氢的催化分解。取两支试管分别加入 2mL w 为 0.03 的 H_2O_2 溶液,将其中一支试管置于水浴中加热,有何现象?用带余烬的火柴放在管口,有何现象?在另一支试管内加入少许 MnO_2 固体,有何现象?迅速用带余烬的火柴放在管口,有何现象?比较以上两种情况,MnO_2 对 H_2O_2 的分解起到什么作用?写出反应方程式。

3. 硫代硫酸盐的性质

a. 硫代硫酸钠与 Cl_2 的反应。取 1mL 0.1mol/L $Na_2S_2O_3$ 溶液于一试管中,加入 2 滴 2mol/L NaOH 溶液,再加入 2mL Cl_2 水,充分振荡,检验溶液中有无 SO_4^{2-} 生成。

b. 硫代硫酸钠与 I_2 的反应。取 1mL 0.1mol/L $Na_2S_2O_3$ 溶液于一试管中,滴加碘水,边滴边振荡,有何现象?此溶液中能否检出 SO_4^{2-}?

c. 硫代硫酸钠的配位反应。取 0.5mL 0.1mol/L $AgNO_3$ 溶液于一试管中,连续滴加 0.1mol/L $Na_2S_2O_3$ 溶液,边滴边振荡,直至生成的沉淀完全溶解。解释所见现象。

4. 离子鉴定

参照本书附录四"常见离子鉴定方法",分别进行 SO_4^{2-}、SO_3^{2-}、$S_2O_3^{2-}$、S^{2-} 的鉴定。写出鉴定的步骤及观察到的现象。

5. 卤酸盐的氧化性

(1) 氯酸钾的氧化性

a. 取少量 $KClO_3$ 晶体置于试管中,用少量水稀释后,加入少许浓盐酸,注意逸出气体的气味,检验气体产物,写出反应式,并作出解释。

b. 分别试验饱和 $KClO_3$ 溶液与 0.1mol/L Na_2SO_3 在中性及酸性条件下的反应,用 $AgNO_3$ 验证反应产物,通过实验说明 $KClO_3$ 的氧化性与介质酸碱性的关系。

c. 取少量 $KClO_3$ 晶体,用 1~2mL 水溶解后,加入少量四氯化碳及 0.1mol/L KI 溶液数滴,摇动试管,观察试管内水相及有机相有什么变化?再加入 6mol/L H_2SO_4 酸化,溶液又有什么变化?写出反应式。能否用 HNO_3 或盐酸来酸化溶液?为什么?

(2) 溴酸钾的氧化性(在通风橱内进行)

a. 饱和溴酸钾溶液经 H_2SO_4 酸化后分别与 0.5mol/L KBr 溶液及 0.5mol/L KI 溶液反应,观察现象并检验反应产物,写出反应式。

b. 试验 $KBrO_3$ 溶液与 Na_2SO_3 溶液在中性及酸性条件下的反应,记录现象,写出反应式。

(3) 碘酸盐的氧化性

0.1mol/L KIO_3 溶液经 3mol/L H_2SO_4 酸化后加入几滴淀粉溶液,再滴加 0.1mol/L Na_2SO_3 溶液,观察现象,写出反应式。若体系不酸化,又有什么现象?改变加入试剂顺序(先加 Na_2SO_3 后滴加 KIO_3),会有什么现象?

6. 金属卤化物的性质

(1) 卤化物的溶解度比较

a. 分别向盛有 0.1mol/L NaF、NaCl、KBr 以及 KI 溶液的试管中滴加 0.1mol/L $Ca(NO_3)_2$ 溶液,观察现象,写出反应式。

b. 分别向盛有 0.1mol/L NaF、NaCl、KBr 以及 KI 溶液的试管中滴加 0.1mol/L $AgNO_3$ 溶液,制得的卤化银沉淀经离心分离后分别与 2mol/L HNO_3,2mol/L $NH_3 \cdot H_2O$ 及 0.5mol/L $Na_2S_2O_3$ 溶液反应,观察沉淀是否溶解?写出反应式。解释氟化物与其卤化物溶解度的差异及变化规律。

（2）卤化银的感光性

将制得的 AgCl 沉淀均匀地涂在滤纸上，滤纸上放一把钥匙，光照约 10min 后取出钥匙，可清楚地看到钥匙的轮廓。卤化银见光分解氯化银较快，碘化银最慢。

7. 小设计

混合溶液中含有 Cl^-、Br^- 和 I^-，试设计实验方案加以鉴别。

四、注意事项

氯酸钾是强氧化剂，保存不当时容易引起爆炸，它与硫、磷的混合物是炸药，因此绝对不允许将它们混在一起。氯酸钾容易分解，不宜大力研磨、烘干或烤干。在进行有关氯酸钾的实验时，如同进行其他有强氧化性物质的实验一样，应将剩余的试剂倒入回收瓶内回收处理，不准倒入废液缸中。

五、思考题

1. 过氧化氢是否既可作为氧化剂又可作为还原剂？什么条件下过氧化氢可将 Mn^{2+} 氧化为 MnO_2？什么条件下 MnO_2 又可将过氧化氢氧化产生氧气？

2. 如何证实亚硫酸盐中存在 SO_4^{2-}？为什么亚硫酸盐中常常有硫酸盐，而硫酸盐中却很少有亚硫酸盐？怎样检验 SO_4^{2-} 盐中的 SO_3^{2-}？

Experiment 8　Identification of Oxygen and Halogen Elements

1. Purpose

(1) To grasp the chemical properties of peroxide and sodium thiosulfate, and master the identification of peroxide.

(2) To master the identification of SO_4^{2-}, SO_3^{2-}, $S_2O_3^{2-}$, S^{2-}.

(3) To learn the oxidizing strength of oxysalts of halogens.

(4) To understand the properties of some metal halides.

2. Apparatus and Chemicals

Apparatus: Small test tubes; water bath and so on.

Chemicals: H_2SO_4 (2mol/L, 6mol/L, 3mol/L); HCl (2mol/L, 6mol/L); NaOH (2mol/L); $Pb(NO_3)_2$ (0.1mol/L); $BaCl_2$ (0.1mol/L); K_2CrO_4 (0.1mol/L); H_2O_2 ($w=0.03$); $AgNO_3$ (0.1mol/L); $K_4[Fe(CN)_6]$ solution (0.1mol/L); $Na_2S_2O_3$ (0.1mol/L); saturated $ZnSO_4$ solution; $MnSO_4$ (0.1mol/L); $Na_2[Fe(CN)_5NO]$ solution; chlorine water; iodine water; thioacetamide solution ($w=0.05$); MnO_2(s); diethyl ether; crystal $KClO_3$; concentrated HCl; $KClO_3$ (saturated solution); Na_2SO_3 (0.1mol/L); CCl_4; KI (0.1mol/L, 0.5mol/L); $KBrO_3$ (saturated solution); KBr (0.1mol/L, 0.5mol/L); NaF (0.1mol/L); NaCl (0.1mol/L); KI (0.1mol/L) $Ca(NO_3)_2$ (0.1mol/L); HNO_3 (2mol/L); $NH_3 \cdot H_2O$ (2mol/L); starch solution and so on.

3. Procedure

3.1 Identification of hydrogen peroxide

Place 2mL of H_2O_2 solution ($w=0.03$) in a test tube, and add 0.5mL diethyl ether and 1mL 2mol/L H_2SO_4. Then add 3-5 drops of 0.1mol/L K_2CrO_4 solution. Observe the color changes of the solution and the diethyl ether layer. Identify the hydrogen peroxide through experiment.

3.2 Properties of hydrogen peroxide

a. Oxidizability of H_2O_2. Mixed drops of 0.1mol/L $Pb(NO_3)_2$ solution and thioacetamide solution ($w=0.05$) in a test tube. Heat it under the water bath. What will happen? After precipitate forms, centrifugalize, separate and decant the above opaque solution. Add a small amount of H_2O_2 ($w=0.03$) to the precipitate which has been washed by distilled water. What will happen to the precipitate? Observe the phenomenon and explain in your report.

b. Reducibility of H_2O_2. Add 2mol/L NaOH solution to 0.5mL 0.1mol/L $AgNO_3$ solution when brown precipitate is produced, then add a small quantity of H_2O_2 ($w=0.03$), observe the phenomenon. Test the produced gas by a match ember. Please explain above phenomena.

c. The Influence of acidity and alkalinity of medium to oxidizability and reducibility of H_2O_2. Add several drops of 2mol/L NaOH solution to 1mL of H_2O_2 ($w=0.03$), then add several drops of 0.1mol/L $MnSO_4$ solution. Shake it and observe the phenomenon. Keep the solution standing, decant the clear solution and add small amount of 2mol/L H_2SO_4 solution to the precipitate. Then H_2O_2 ($w=0.03$) is added, what will you observe? Write down the reaction equation and explain it.

d. Catalytic decomposition of H_2O_2. Heat 2mL H_2O_2 solution ($w=0.03$) under water bath, and test the produced gas by match ember. What will happen? Mixed 2mL H_2O_2 solution ($w=0.03$) and a small quantity of solid MnO_2, then test the produced gas by match ember. Observe the phenomenon. Summarize the two experiments and explain the effect of MnO_2 in the reaction of decomposition of H_2O_2. Write down the reaction equation.

3.3 Properties of thiosulfate

a. Reaction of $Na_2S_2O_3$ and Cl_2. Place 1 mL 0.1mol/L $Na_2S_2O_3$ solution, two drops of 2mol/L NaOH solution and 2mL chlorine water in a test tube. Shake it and test whether there is SO_4^{2-}.

b. Reaction of $Na_2S_2O_3$ and I_2. Add iodine water to 1 mL 0.1mol/L $Na_2S_2O_3$ solution in a test tube. What will happen? Is there SO_4^{2-} in the solution?

c. Coordination reaction of $Na_2S_2O_3$. Place 0.5mL 0.1mol/L $AgNO_3$ solution in a test tube, and add $Na_2S_2O_3$ solution drop by drop until the produced precipitate is dissolved. Record your observations.

3.4 Identification of SO_4^{2-}, SO_3^{2-}, $S_2O_3^{2-}$ and S^{2-}

Identify SO_4^{2-}, SO_3^{2-}, $S_2O_3^{2-}$ and S^{2-} respectively according to the identification method listed in appendix 4. Record your observations and the identification steps.

3.5 Oxidizability of chlorate

(1) Oxidizability of potassium chlorate

a. Add a small quantity of distilled water to crystal $KClO_3$, then a small quantity of concentrated hydrochloric acid is added. Observe the gas produced, test the gas, write down the reaction equations, and explain the phenomenon.

b. Test the reaction between saturated $KClO_3$ and 0.1mol/L Na_2SO_3 solution under neutral and acid medium. Test the product produced by $AgNO_3$. What conclusion can you draw on the relationship between the oxidizability of potassium chlorate and acid or alkaline properties of medium?

c. Fetch a small quantity of crystal $KClO_3$, dissolve it with 1-2mL water, add a small quantity of CCl_4 and 0.1mol/L KI solution, shake the test tube, observe the phenomena of organic phase and water phase. Add 6mol/L H_2SO_4 solution, what happens to it? Write down the reaction equation. Can you acidify the solution with HNO_3? Why?

(2) Oxidizability of potassium bromate (carried out in fume hood)

a. Acidify the saturated potassium bromate solution by H_2SO_4 in two test tubes, then add 0.5mol/L KBr solution and 0.5mol/L KI solution respectively, observe and test the product. Write down the reaction equations.

b. Test the reaction between $KBrO_3$ solution and Na_2SO_3 solution in neutral and acid medium. Record the phenomenon and write down the reaction equation.

(3) Oxidizability of iodate

0.1mol/L KIO_3 solution acidified with 3mol/L H_2SO_4 is added to several drops of starch solution, then add 0.1mol/L Na_2SO_3 solution. Observe the phenomenon and write down the reaction equation. If the mixture isn't acidified, what phenomenon will we observe? Change the sequence when you add the reagents. What phenomenon will be observed?

3.6 The properties of metal halide

(1) Comparison of halide's solubility

a. Drop 0.1mol/L $Ca(NO_3)_2$ solution to 0.1mol/L NaF, NaCl, KBr and KI respectively. Observe the phenomenon, write down the reaction equations.

b. Drop 0.1mol/L $AgNO_3$ solution to the test tubes with 0.1mol/L NaF, NaCl, KBr and KI respectively. Centrifugalize the precipitate, then let precipitate react with 2mol/L HNO_3, 2mol/L $NH_3 \cdot H_2O$ and 0.5mol/L $Na_2S_2O_3$ solution, observe the dissolving of precipitate. Write down the reaction equations. Explain the differences of solubility between fluoride and their halides and summarize the changing law.

(2) Light sensitiveness of silver halides

Daub produced AgCl on filter paper, put a key on the filter paper, illuminate it about ten minutes, clear adumbration can be seen. AgCl decomposes most quickly, while AgI does slowly.

3.7 Design

For mixture containing Cl^-, Br^- and I^-, design a separation and identification scheme.

4. Notes

Potassium chlorate is a strong oxidizing agent, improper preservation may cause explosion. The mixture of potassium chlorate, sulfur and phosphor is dynamite. So you should never mix them together. $KClO_3$ decomposes easily, don't grind emphatically or over dry. Carrying out the experiments with $KClO_3$, as you do experiment with other regents of strong oxidizability, pour surplus reagent into recovery bottle, don't pour it to acid jar.

5. Questions

(1) Can H_2O_2 be used as oxidant and reductant? What is the medium needed if MnO_2 oxidizes H_2O_2 to produce O_2? What is the medium needed if H_2O_2 oxidizes Mn^{2+} to MnO_2?

(2) How to confirm the existence of SO_4^{2-} in sulfite? There always is sulfite in sulfate, but sulfate always does not contain sulfite, why? How to detect SO_3^{2-} in sulfate?

(3) When identifying NO_3^-, how do we remove NO_2^-?

实验九 铬、锰、铁、钴、铜、银、锌、汞元素鉴定

一、实验目的

1. 了解各元素的性质。
2. 掌握 Cr^{3+}、Mn^{2+}、Fe^{3+}、Cu^{2+} 等离子的鉴定方法。

二、仪器与药品

仪器：离心机；小试管；水浴锅；点滴板等。

药品：$AgNO_3$（0.1mol/L、0.5mol/L）；H_2SO_4（1mol/L）；$CrCl_3$（0.1mol/L）；$K_2Cr_2O_7$（0.1mol/L）；NaOH（2mol/L，w 为 0.4，6mol/L）；$BaCl_2$（0.1mol/L）；$MnCl_2$（0.1mol/L）；$MnSO_4$（0.1mol/L）；浓盐酸；$KMnO_4$（0.01mol/L）；MnO_2(s)；$Co(NO_3)_2$（0.1mol/L）；HCl（2mol/L）；$K_4[Fe(CN)_6]$ 溶液；Fe^{2+} 溶液；Fe^{3+} 溶液；$K_3[Fe(CN)_6]$ 溶液；$CuSO_4$（0.1mol/L）；w 为 0.10 的葡萄糖溶液；NaCl(s)；铜粉；浓氨水；浓硫酸；$ZnSO_4$（0.1mol/L）；$HgCl_2$（0.1mol/L）；$SnCl_2$（0.1mol/L）；$(NH_4)_2Fe(SO_4)_2$(s)；$CuCl_2$（1mol/L）；KI（0.1mol/L）；$AgNO_3$（0.1mol/L）；H_2O_2（w 为 0.03）；HAc（6mol/L）；$Pb(NO_3)_2$（6mol/L）；$NH_3 \cdot H_2O$（2mol/L）；$Hg_2(NO_3)_2$（0.1mol/L）。

三、实验步骤

1. 铬的化合物

（1）氢氧化铬的生成与性质

以 0.1mol/L $CrCl_3$ 溶液为原料，自行设计实验制备 $Cr(OH)_3$，并试验 $Cr(OH)_3$ 是否具有两性。

(2) Cr^{3+} 的氧化

以 0.1mol/L $CrCl_3$ 溶液为原料，自行设计实验将其氧化为 CrO_4^{2-}，并写出反应方程式。设计实验时要注意：

a. Cr^{3+} 的氧化宜在较强的碱性介质中进行；

b. 该氧化反应的反应速率较慢，宜加热进行。

(3) $Cr_2O_7^{2-}$ 和 CrO_4^{2-} 的相互转化

在 0.5mL 0.1mol/L $K_2Cr_2O_7$ 的溶液中，滴入少许 2mol/L NaOH 溶液，观察溶液颜色变化。然后加入 1mol/L H_2SO_4 酸化，观察溶液颜色又有何变化？解释现象，并写出 $Cr_2O_7^{2-}$ 与 CrO_4^{2-} 之间的平衡方程式。

在 5 滴 0.1mol/L $K_2Cr_2O_7$ 的溶液中，加入数滴 0.1mol/L $BaCl_2$ 溶液，有何现象产生？为什么得到的沉淀不是 $BaCr_2O_7$？写出反应方程式。

(4) Cr^{3+} 的鉴定

Cr^{3+} 的鉴定请参看本书附录四，然后按照所列步骤鉴定 Cr^{3+}。

2. 锰的化合物

(1) $Mn(OH)_2$ 的生成和性质

以 0.1mol/L $MnCl_2$ 溶液为原料，自行设计实验制备 $Mn(OH)_2$，并试验 $Mn(OH)_2$ 是否具有两性。

把制得的一部分 $Mn(OH)_2$ 沉淀在空气中放置一段时间，注意沉淀颜色的变化，并解释之。写出反应方程式。

(2) MnO_4^{2-} 的生成

在盛有 2mL 0.01mol/L $KMnO_4$ 溶液的试管中，加入 1mL w 为 0.40 的 NaOH 溶液，然后加入少量 MnO_2 固体，微热，不断搅动 2min。静置片刻，待 MnO_2 沉降后观察上层溶液的颜色，写出反应方程式。

取出部分上层清液，加入 1mol/L H_2SO_4 酸化，观察溶液颜色的变化和沉淀的生成。写出反应方程式。通过以上实验，说明 MnO_4^{2-} 存在的条件。

(3) Mn(Ⅶ) 和 Mn(Ⅳ) 氧化性的比较

用 0.01mol/L $KMnO_4$ 溶液和固体 MnO_2，分别与浓 HCl 溶液、0.1mol/L $MnSO_4$ 溶液反应。根据实验结果，比较 Mn(Ⅶ) 和 Mn(Ⅳ) 氧化性的强弱。写出反应方程式。

3. 铁和钴的化合物

(1) $Fe(OH)_2$ 的生成和性质

在试管中加入 1mL 去离子水，煮沸后赶尽空气。待其冷却后，再加入 2 滴浓硫酸和一小粒 $(NH_4)_2Fe(SO_4)_2$ 固体，用玻璃棒轻轻搅动使其溶解。

在另一支试管中加入 1mL 6mol/L NaOH 溶液，煮沸后赶尽空气。冷却后，用滴管吸取 0.5mL 溶液，插入上述盛有 $FeSO_4$ 溶液的试管底部，慢慢放出 NaOH 溶液（整个操作要避免将空气带入溶液）。观察所生成沉淀的颜色。放置一段时间后，观察沉淀颜色有何变化。写出反应方程式。

(2) $Co(OH)_2$ 的生成和性质

在 5 滴 0.1mol/L $Co(NO_3)_2$ 溶液中滴加 2mol/L NaOH 溶液，观察所生成沉淀的颜色。微热，沉淀的颜色有何变化？放置一段时间后，沉淀的颜色又有何变化？写出反应方程式。

4. Fe^{2+} 和 Fe^{3+} 的鉴定

(1) Fe^{3+} 的鉴定

分别取待鉴定溶液 1 滴，放在点滴板上，向各溶液加 1 滴 2mol/L HCl 及 1 滴 $K_4[Fe(CN)_6]$ 溶液，观察生成沉淀的情况和颜色，记录现象。

(2) Fe^{2+} 的鉴定

分别取待鉴定溶液 1 滴，放在点滴板上，向各溶液加 1 滴 2mol/L HCl 及 1 滴 $K_3[Fe(CN)_6]$ 溶液，观察生成沉淀的情况和颜色，记录现象。

用方程式解释上述实验现象，并得出结论。

5. 铜的化合物

(1) Cu_2O 的生成

在试管中加入 0.5mL 0.1mol/L $CuSO_4$ 溶液，加入过量的 6mol/L NaOH 至初生成的沉淀完全溶解，再往此溶液中加入 0.5mL w 为 0.10 的葡萄糖溶液。混匀后微热之。观察现象。

(2) CuCl 的生成和性质

取 5mL 1mol/L $CuCl_2$ 溶液，加少量固体 NaCl 和铜粉，加热至沸，当溶液变成棕黄色时，将溶液迅速倒入盛有 20mL 水的烧杯中，充分搅拌后有何现象？若有沉淀，静置让其沉淀，用倾析法倾出溶液，将沉淀分为两份，在沉淀中分别加入浓氨水和浓盐酸，观察现象，写出反应方程式。

(3) CuI 的生成

往 5 滴 0.1mol/L $CuSO_4$ 溶液中，滴加 0.1mol/L KI 溶液，观察现象。为了消除 I_2 的颜色干扰，再往溶液中滴加 0.1mol/L $Na_2S_2O_3$ 溶液至 I_2 的棕色褪去，观察沉淀的颜色。写出反应方程式。

6. 银的化合物

(1) Ag_2O 的生成和性质

在盛有 0.5mL 0.1mol/L 的 $AgNO_3$ 溶液的离心管中，慢慢滴加新配制的 2mol/L NaOH 溶液。观察生成的 Ag_2O 沉淀的颜色。离心分离，弃去溶液，用去离子水洗涤沉淀。将沉淀分为两份分别检测其酸碱性。

(2) 银镜的制备

向一洁净的试管中加入 1mL 0.1mol/L 的 $AgNO_3$ 溶液，滴加 2mol/L $NH_3 \cdot H_2O$ 至生成的沉淀完全溶解。随后加入数滴 w 为 0.1 的葡萄糖溶液，在水浴上加热，观察试管壁有何变化，写出反应方程式。

7. 锌和汞的化合物

(1) $Zn(OH)_2$ 的生成和性质

自行设计实验制备 $Zn(OH)_2$，并检测其酸碱性。写出有关反应方程式。

(2) Hg(Ⅱ) 和 Hg(Ⅰ) 的相互转化

在 0.5mL 0.1mol/L $HgCl_2$ 溶液中，逐滴加入 0.1mol/L $SnCl_2$ 溶液，边加边振荡，注意沉淀颜色的变化过程。写出反应方程式。

在 0.5ml 0.1mol/L $Hg_2(NO_3)_2$ 溶液中，逐滴加入 2mol/L 的氨水，振荡，观察沉淀的颜色并写出反应方程式。

四、思考题

1. 试分析为什么 $CuSO_4$ 中加入 KI 会产生 CuI 沉淀，而加入 KCl 时却不出现 CuCl 沉淀？
2. Hg 和 Hg^{2+} 有剧毒，试验时应注意什么？
3. 制备 $Fe(OH)_2$ 时，有关溶液均需煮沸并避免振荡，为什么？

Experiment 9 Identification of Chromium, Manganese, Iron, Cobalt, Copper, Silver, Zinc and Mercury

1. Purpose

(1) To learn the properties of these elements.

(2) To master the identification methods of Cr^{3+}, Mn^{2+}, Fe^{3+}, Cu^{2+} ions and so on.

2. Apparatus and Chemicals

Apparatus: Centrifuge; small test tubes; water bath; spot plate and so on.

Chemicals: $AgNO_3$ (0.1mol/L, 0.5mol/L); H_2SO_4 (1mol/L); $CrCl_3$ (0.1mol/L); $K_2Cr_2O_7$ (0.1mol/L); NaOH (2mol/L, $w=0.4$, 6mol/L); $BaCl_2$ (0.1mol/L); $MnCl_2$ (0.1mol/L); $MnSO_4$ (0.1mol/L); concentrated HCl; $KMnO_4$ (0.01mol/L); MnO_2 (s); $Co(NO_3)_2$ (0.1mol/L); HCl (2mol/L); $K_4[Fe(CN)_6]$ solution; Fe^{2+} solution; Fe^{3+} solution; $K_3[Fe(CN)_6]$ solution; $CuSO_4$ (0.1mol/L); glucose solution ($w=0.10$); NaCl (s); copper powder; concentrated ammonia water; concentrated H_2SO_4; $ZnSO_4$ (0.1mol/L); $HgCl_2$ (0.1mol/L); $SnCl_2$ (0.1mol/L); $(NH_4)_2Fe(SO_4)_2$ (s); $CuCl_2$ (1mol/L); KI (0.1mol/L); $AgNO_3$ (0.1mol/L); H_2O_2 ($w=0.03$); HAc (6mol/L); $Pb(NO_3)_2$ (6mol/L); $NH_3 \cdot H_2O$ (2mol/L); $Hg_2(NO_3)_2$ (0.1mol/L).

3. Procedure

3.1 Compounds of Cr

(1) The property, formation of $Cr(OH)_3$ (Ⅲ)

Check relevant information and design experimental scheme to prepare $Cr(OH)_3$ with 0.1mol/L $CrCl_3$ solution. Test the amphoteric property of $Cr(OH)_3$.

(2) Reducibility of Cr^{3+}

Design experimental scheme to oxidize the Cr^{3+} to CrO_4^{2-} with 0.1mol/L $CrCl_3$ solution. And write down the reaction equations. The following aspect should be noted:

a. It is easy for Cr^{3+} to be oxidized under the strong basic condition.

b. It is better to heat them during the experiment because of the low reaction rate.

(3) The transformation of $Cr_2O_7^{2-}$ and CrO_4^{2-}

Add two drops of 2mol/L NaOH to 0.5mL 0.1mol/L $K_2Cr_2O_7$ solution and observe the phenomena.

Add small quantities of $BaCl_2$ solution into 5 drops of 0.1mol/L $K_2Cr_2O_7$ solution. Observe the phenomena and think about the reason why the precipitate is not $BaCr_2O_7$. Write down the reaction equations.

(4) Identification of Cr^{3+}

Please refer to the appendix 4 of this book fot the method of identifying Cr^{3+}.

3.2　Compounds of Mn

(1) The formation and property of $Mn(OH)_2$

Design experimental scheme to prepare $Mn(OH)_2$ with 0.1mol/L $MnCl_2$ solution. Test whether the $Mn(OH)_2$ is amphoteric property.

Put a portion of the $Mn(OH)_2$ precipitate in the air for a period of time. Observe the change of the color and explain the reason. Then write down the reaction equations.

(2) The formation of MnO_4^{2-}

Decant the above clear liquid. Heat gently and stir it for 2 min. Let it stand for a while. Observe the color of the solution and write down the reaction equations.

Decant above the layer and acidify it with 1mol/L H_2SO_4, and observe the color change and the formation of the precipitate. Write down the reaction equations and summarize the existing condition of MnO_4^{2-}.

(3) Compare the oxidation of Mn(Ⅶ) and Mn(Ⅳ)

Place a small amount of 0.01mol/L $KMnO_4$ solution and solid MnO_2 to react with concentrated HCl and 0.1mol/L $MnSO_4$ respectively. Compare the oxidation of Mn(Ⅶ) and Mn(Ⅳ) according to the result of the experiment and write down the reaction equations.

3.3　Compound of Fe and Co

(1) The formation and property of $Fe(OH)_2$

Add 1mL distilled water into a test tube, heat to boil in order to drive off the air, then add $(NH_4)_2Fe(SO_4)_2$ solid of a mung bean size and two drops of concentrated H_2SO_4 when it is cold. Stir briskly to dissolve the solid.

Add 1mL 6mol/L NaOH into another test tube, heat to boil in order to drive off the air. Take 0.5mL NaOH solution with a long burette after cooling down, then insert it into $(NH_4)_2Fe(SO_4)_2$ solution of the former test tube (direct to bottom of test tube), slowly drop NaOH (preventing air from entering the solution in the whole procedure). Observe the color and state of product. Let the sample stand for a while after shaking the tube, and observe the change. Write down the related reaction equation.

(2) The formation and property of $Co(OH)_2$

Dropping 2mol/L NaOH solution to 5 drops of 0.1mol/L $Co(NO_3)_2$, observe the color of precipitate. Heat it gently and observe the color change of the product. Let the sample stand for a while, and observe the change. Write down the related reaction equation.

3.4　Identification of Fe^{2+} and Fe^{3+}

(1) Identification of Fe^{3+}

Add a drop of Fe^{3+} solution on two spot plates respectively. Then add one drop 2mol/L

HCl and one drop $K_4[Fe(CN)_6]$ solution. Observe the color and state of the precipitate.

(2) Identification of Fe^{2+}

Add a drop of Fe^{3+} solution on two spot plates respectively. Then add one drop 2mol/L HCl and one drop $K_3[Fe(CN)_6]$ solution. Observe the color and state of the precipitate.

Sum up the principle and write down the reaction equation on the basis of above experimental phenomena.

3.5 Compound of Cu

(1) The formation of Cu_2O

Add excessive 6mol/L NaOH to a 0.5mL 0.1mol/L $CuSO_4$ solution. After the produced precipitate is dissolved, add 0.5mL glucose solution ($w=0.1$). Mix well and heat gently. Observe the phenomena.

(2) The formation and property of CuCl

Add 5mL 1mol/L $CuCl_2$ solution to a test tube, then fetch small quantities of solid NaCl and copper powder in it. Heat to boil, if it appears dark brown, pour all the solution to 20mL distilled water rapidly, observe the phenomena. When bulk of precipitate appears, let it stand, decant above the layer, and divide the precipitate into two portions. One reacts with concentrated ammonia water, another with concentration HCl solution. Whether does precipitate dissolve? Write down the reaction equations.

(3) The formation and property of CuI

Add 5 drops of 0.1mol/L $CuSO_4$ solution to a test tube, then add 0.1mol/L KI solution drop by drop. What will you observe? Drops of 0.1mol/L $Na_2S_2O_3$ solution are added to eliminate the interference of the color of I_2 until the color of I_2 has died away. Observe the phenomena and write down the reaction equations.

3.6 Compound of Ag

(1) The formation and property of Ag_2O

Add 0.5mL 0.1mol/L $AgNO_3$ solution to the tube, 2mol/L NaOH solution is then added dropwise, shake, observe the color and state of Ag_2O. Centrifugalize the tube, remove solution of above layer. Wash the precipitate with distilled water. Divide the precipitate into two portions. One reacts with acid solution, the other with base solution. Observe the phenomena and write down the reaction equations.

(2) Silver mirror reaction

1mL 0.1mol/L $AgNO_3$ is added to one clean test tube, 2mol/L $NH_3 \cdot H_2O$ is added until precipitate is dissolved. Then drops of 10% glucose solution are added, shake and place it in water bath. What will be seen in the test tube? Write down the reaction equations.

3.7 Compounds of Zn and Hg

(1) The formation and property of $Zn(OH)_2$

Design experimental scheme to prepare $Zn(OH)_2$ and test whether the solution is acidic or alkaline.

(2) Transformation of Hg(II) and Hg(I)

Add 0.1mol/L $SnCl_2$ dropwise to 0.5mL 0.1mol/L $HgCl_2$ solution, shake it. Observe the phenomena and write down the reaction equations.

Add 2mol/L $NH_3·H_2O$ dropwise to 0.5mL 0.1mol/L $Hg_2(NO_3)_2$ solution and shake it. Observe the color of the precipitate and write down the reaction equations

4. Questions

(1) Adding KI to $CuSO_4$ solution can form CuI precipitate, but adding KCl to $CuSO_4$ solution can not form CuCl precipitate, Why?

(2) Hg and Hg^{2+} are toxic. What precautions should you take when doing experiments with them?

(3) When preparing $Fe(OH)_2$, the solutions should be boiled and shaking should be avoided, Why?

实验十 盐酸标准溶液的配制与标定

一、实验目的

1. 掌握盐酸溶液的配制方法。
2. 掌握标定盐酸溶液浓度的基本原理和基本方法。
3. 掌握酸式滴定管的使用方法和酸碱滴定的基本操作。

二、实验原理

盐酸容易挥发，其标准溶液不能直接配制，需要先配制成近似浓度，然后用基准物质标定其准确浓度，常以无水碳酸钠作为基准物质来标定盐酸标准溶液。方程式如下：

$$Na_2CO_3 + 2HCl = 2NaCl + H_2O + CO_2\uparrow$$

选用溴甲酚绿-二甲基黄混合指示剂，滴定终点颜色由绿色变为亮黄色（pH=3.9），根据 Na_2CO_3 的质量和所消耗的 HCl 的体积，按照下式计算 HCl 的物质的量浓度。

$$c_{HCl} = \frac{2w_{Na_2CO_3}(g) \times 1000}{V_{HCl} \times M_{Na_2CO_3}}$$

$$M_{Na_2CO_3} = 105.99 g/mol$$

三、仪器与药品

仪器：分析天平；25mL 酸式滴定管；250mL 锥形瓶；100mL 量筒；25mL 移液管；250mL 容量瓶。

药品：HCl（36%～38%，相对密度为1.18）；Na_2CO_3（270～290℃）；溴甲酚绿-二甲基黄混合指示剂。

四、实验步骤

1. 0.1mol/L 盐酸溶液的配制

计算配制 500mL 0.1mol/L 盐酸溶液所需浓盐酸的体积。量取计算体积的浓盐酸，倒入盛有适量去离子水的试剂瓶中，加水稀释至500mL，摇匀。

2. 盐酸溶液浓度的标定

用分析天平准确称取0.10～0.12g无水Na_2CO_3，置于250mL锥形瓶中，加入80mL去离子水搅拌，使Na_2CO_3完全溶解。加入9滴溴甲酚绿-二甲基黄混合指示剂，用滴定管慢慢滴入待测盐酸溶液，当锥形瓶中溶液由蓝绿色变为亮黄色即为滴定终点。记下滴定消耗HCl的体积（mL）。平行测定三次。

五、数据记录及结果处理

数据记录及结果计算见表3-4。

表3-4 数据记录及结果计算

试剂	1	2	3
$M_{Na_2CO_3}$/g			
V_{HCl}/mL			
c_{HCl}/(mol/L)			
$\bar{c}_{HCl}$/(mol/L)			
相对平均偏差/%			

六、注意事项

1. 因碳酸钠容易吸水，称量要迅速。
2. 实验中使用的锥形瓶不需要干燥。

七、思考题

1. 用碳酸钠为基准物标定盐酸溶液（0.2mol/L）时，基准物的取量如何计算？
2. 用碳酸钠标定盐酸溶液时为什么需要用溴甲酚绿-二甲基黄混合指示剂？
3. 0.07980的有效数字为几位？

Experiment 10 Preparation and Standardization of Hydrochloride Acid Standard Solution

1. Purpose

(1) To master the preparation method of hydrochloric acid solution.

(2) To master the basic principles and methods of calibrating the concentration of hydrochloric acid solution.

(3) To master the usage method of acid burette and the basic operation of acid-base titration.

2. Principle

Hydrochloric acid is prone to volatilization, and its standard solution cannot be directly prepared. It is necessary to first prepare an approximate concentration and then calibrate its accurate concentration with a reference substance. Anhydrous sodium carbonate is often used as the reference substance to calibrate the hydrochloric acid standard solution. The equation is as follows:

$$Na_2CO_3 + 2HCl = 2NaCl + H_2O + CO_2 \uparrow$$

Using a mixed indicator of bromocresol green-dimethyl yellow, the endpoint color of the titration changes from green to bright yellow (pH = 3.9). Based on the mass of Na_2CO_3 and the volume of HCl consumed, calculate the concentration of HCl according to the following formula.

$$c_{HCl} = \frac{2w_{Na_2CO_3}(g) \times 1000}{V_{HCl} \times M_{Na_2CO_3}}$$

$$M_{Na_2CO_3} = 105.99 \text{g/mol}$$

3. Apparatus and Chemicals

Apparatus: 25mL acid burette; 100mL volumetric cylinder; 250mL conical flask; 25mL transferring pipets.

Chemicals: Na_2CO_3 (primary standard reagent); HCl (36%-38%, relative density is 1.18); bromocresol green-two methyl yellow mixed indicator.

4. Procedure

(1) Preparation of 0.1mol/L hydrochloride acid solution

Calculate the volume of concentrated hydrochloric acid required to prepare 500mL of 0.1mol/L hydrochloric acid solution. Measure the volume of concentrated hydrochloric acid and pour it into a reagent bottle containing an appropriate amount of deionized water. Dilute with water to 500mL and shake well.

(2) Calibration of hydrochloric acid solution concentration

Accurately weigh 0.10-0.12g of anhydrous Na_2CO_3 using an analytical balance and place it in a 250mL conical flask. Add 80mL of deionized water and stir until Na_2CO_3 is completely dissolved. Add 9 drops of bromocresol green-dimethyl yellow mixed indicator, and slowly drop it into the hydrochloric acid solution to be tested using a burette. The point when the solution in the conical flask changes from blue-green to bright yellow is the titration endpoint. Record the volume (mL) of HCl consumed during titration. Measure three times in parallel.

5. Data Analysis

Table 3-4 is data record and analysis.

Table 3-4 Data record and analysis

Reagents	1	2	3
$M_{Na_2CO_3}$/g			
V_{HCl}/mL			
c_{HCl}/(mol/L)			
$\bar{c}_{HCl}$/(mol/L)			
Relative average deviation/%			

6. Notes

(1) As sodium carbonate easily absorbs water, it needs to be weighed quickly.

(2) The conical flask used in the experiment does not need to be dried.

7. Questions

(1) How to calculate the amount of reference substance when calibrating hydrochloric acid solution (0.2mol/L) with sodium carbonate as the reference substance?

(2) Why is it necessary to use a mixed indicator of bromocresol green-dimethyl yellow when calibrating hydrochloric acid solution with sodium carbonate?

(3) How many significant digits in 0.07980?

实验十一 混合碱中各组分含量的测定（双指示剂法）

一、实验目的

1. 掌握用双指示剂法测定混合碱中 Na_2CO_3、$NaHCO_3$ 和 NaOH 含量的方法。
2. 了解酸碱滴定中强碱弱酸盐的测定原理。
3. 练习移液管、酸式滴定管的基本操作方法。

二、实验原理

混合碱是指 Na_2CO_3 与 NaOH 或 Na_2CO_3 与 $NaHCO_3$ 的混合物，一般可以采用双指示剂法进行测定。双指示剂法是利用指示剂在不同化学计量点的颜色变化，得到两个终点，分别根据各终点所消耗的酸标准溶液的体积，计算各组分的含量。

以 Na_2CO_3 和 NaOH 组成的混合碱为例。实验中首先加入酚酞指示剂，以 HCl 标准溶液滴定至溶液刚好褪色，此时溶液中的 NaOH 被完全中和，Na_2CO_3 仅被滴定成 $NaHCO_3$，反应式如下：

$$NaOH + HCl = NaCl + H_2O$$

$$Na_2CO_3 + HCl = NaHCO_3 + NaCl$$

此时消耗的盐酸体积为 V_1 mL。随后向溶液中再加入 9 滴溴甲酚绿-二甲基黄指示剂，继续用 HCl 滴定至溶液由蓝绿色变为亮黄色，即为滴定终点。此时溶液中的 $NaHCO_3$ 完全被中和，有关的反应为：

$$NaHCO_3 + HCl \longrightarrow NaCl + CO_2 \uparrow + H_2O$$

设此时消耗的盐酸体积为 V_2 mL。

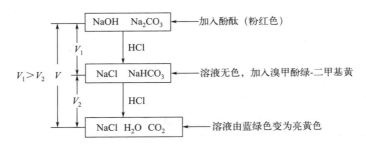

由上述反应可知，第一次变色，NaOH 和 Na_2CO_3 共消耗 HCl 标准溶液的体积为 V_1 mL；第二次变色，$NaHCO_3$ 消耗 HCl 标准溶液的体积为 V_2 mL。$V_1 > V_2$，由此推断 NaOH 消耗 HCl 标准溶液的体积为 $(V_1 - V_2)$ mL，Na_2CO_3 消耗 HCl 标准溶液的体积为 $2V_2$ mL。混合碱中 Na_2CO_3 与 NaOH 的含量（mol/L）可由下式计算：

$$c_{NaOH} = \frac{c_{HCl}(V_1 - V_2)}{V_{混合碱}}$$

$$c_{Na_2CO_3} = \frac{c_{HCl}V_2}{V_{混合碱}}$$

根据 V_1 和 V_2 体积的大小，可以推断混合碱的组成及各组分的含量。

三、仪器与药品

仪器：25mL 移液管；25mL 酸式滴定管；250mL 锥形瓶；100mL 烧杯；100mL 量筒。

药品：混合碱试样 1；混合碱试样 2；酚酞指示剂；溴甲酚绿-二甲基黄指示剂；HCl 标准溶液。

四、实验步骤

准确移取 25.00mL 混合碱试样 1，置于 250mL 锥形瓶中，加入 50mL 去离子水混合均匀，滴入 1～2 滴酚酞指示剂，摇匀后用 HCl 标准溶液滴定，滴定至溶液由粉红色至无色，即为滴定终点，记下所用 HCl 标准溶液的体积 V_1。随后向溶液中再加入 9 滴溴甲酚绿-二甲基黄指示剂，继续用 HCl 标准溶液滴定至溶液由蓝绿色变为亮黄色，即为滴定终点，记下所用 HCl 标准溶液的体积 V_2。根据 V_1 和 V_2 判断碱样组成并计算各组分的含量。平行测定三次。碱样 2 重复上述操作。

五、数据记录及结果处理

数据记录及结果处理见表 3-5、表 3-6。

表 3-5 碱样 1 数据记录及结果计算

试剂		1	2	3
酚酞指示剂	V_1/mL			
溴甲酚绿-二甲基黄指示剂	V_2/mL			
碱样 1 的组成				
碱样 1 中组分 1 的含量/(mol/L)				
平均值/(mol/L)				
碱样 1 中组分 2 的含量/(mol/L)				
平均值/(mol/L)				

表 3-6 碱样 2 数据记录及结果计算

试剂		1	2	3
酚酞指示剂	V_1/mL			
溴甲酚绿-二甲基黄指示剂	V_2/mL			
碱样 2 的组成				
碱样 2 中组分 1 的含量/(mol/L)				
平均值/(mol/L)				
碱样 2 中组分 2 的含量/(mol/L)				
平均值/(mol/L)				

六、注意事项

1. 当混合碱由 Na_2CO_3 与 NaOH 组成时，酚酞用量可以适当多加几滴，否则常因滴定不完全而使 NaOH 的测定结果偏低。

2. 第一计量点时，滴定速度要适中，摇动要均匀，避免局部 HCl 过量，生成 H_2CO_3。

七、思考题

1. 本实验是否可以选择其他指示剂？

2. 请判断下列五种情况时，样品溶液的组成：(1) $V_1=V_2$，V_1、$V_2>0$；(2) $V_1=0$，$V_2>0$；(3) $V_1>0$，$V_2=0$；(4) $V_1>V_2$；(5) $V_1<V_2$。

Experiment 11　Determination of the Composition of Mixed Base (Double-tracer Technique)

1. Purpose

(1) To master the dual indicator method to determine the content of Na_2CO_3, $NaHCO_3$, and NaOH in mixed alkali.

(2) To understand the principle of measuring strong base and weak acid salts in acid-base titration.

(3) To practice the basic operating methods of pipette and acid burette.

2. Principle

Mixed alkali refers to the mixture of Na_2CO_3 and $NaOH$ or Na_2CO_3 and $NaHCO_3$, which can generally be determined using the "dual indicator" method. The dual indicator method utilizes the color changes of indicators at different stoichiometric points to obtain two endpoints, and calculate the content of each component based on the volume of acid standard solution consumed at each endpoint.

Taking a mixed alkali composed of Na_2CO_3 and $NaOH$ as an example. In the experiment, phenolphthalein indicator is first added and titrated with HCl standard solution until the color just fades. At this point, NaOH in the solution is completely neutralized, and Na_2CO_3 is only titrated into $NaHCO_3$. The reaction formula is as follows:

$$NaOH + HCl = NaCl + H_2O$$

$$Na_2CO_3 + HCl = NaHCO_3 + NaCl$$

The volume of hydrochloric acid consumed at this time is V_1 mL. Then add 9 drops of bromocresol green-dimethyl yellow indicator to the solution and continue titrating with HCl until the solution changes from blue-green to bright yellow, which is the titration endpoint. At this point, $NaHCO_3$ in the solution is completely neutralized, and the relevant reactions is

$$NaHCO_3 + HCl = NaCl + CO_2 \uparrow + H_2O$$

and the volume of HCl consumed at this time be V_2 mL.

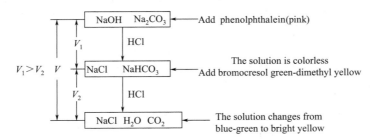

From the above reaction, it can be seen that $V_1 > V_2$, Na_2CO_3 consumes a volume of 2V_2 mL of HCl standard solution, and NaOH consumes a volume of $(V_1 - V_2)$ mL of HCl standard solution. The content of Na_2CO_3 and NaOH in the mixed alkali (mol/L) can be calculated by the following equation:

$$c_{NaOH} = \frac{c_{HCl}(V_1 - V_2)}{V_{Mixed\ alkali}}$$

$$c_{Na_2CO_3} = \frac{c_{HCl} \times V_2}{V_{Mixed\ alkali}}$$

Based on the volume of V_1 and V_2, the composition and content of each component of the mixed alkali can be inferred.

3. Apparatus and Chemicals

Apparatus: 25mL pipette; 25mL acid burette; 250mL conical flask; beaker; 100mL graduated cylinder.

Chemicals: Mixed alkali sample 1; mixed alkali sample 2; phenolphthalein indicator; bromocresol green-dimethyl yellow indicator; HCl standard solution.

4. Procedure

Pipet 25.00mL of mixed base 1 to a 250mL erlenmeyer flask, add 50mL distilled water to it. Add 1-2 drops of phenolphthalein indicator. Titrate with HCl standard solution and judge the endpoint by the color changing from red to colorless. Write down the data as V_1. Add 9 drops of bromocresol green-two methyl yellow in the above solution. Titrate with HCl solution until the color turns from green to bright yellow. Write down the data as V_2. The percentage composition of mixed base can be determinated according to V_1 and V_2. Repeat twice. The determination of mixed base 2 is the same as above.

5. Data Analysis

Data record and analysis are shown in Table 3-5 and Table 3-6.

Table 3-5   Mixed alkali 1: data record and analysis

Reagents		1	2	3
Phenolphthalein indicator	V_1/mL			
Bromocresol green-two methyl yellow indicator	V_2/mL			
Composition of alkali sample 1				
Content of component 1/(mol/L)				
Average value/(mol/L)				
Content of component 2/(mol/L)				
Average value/(mol/L)				

Table 3-6   Mixed alkali 2: data record and analysis

Reagents		1	2	3
Phenolphthalein indicator	V_1/mL			
Bromocresol green-two methyl yellow indicator	V_2/mL			

续表

Reagents	1	2	3
Composition of alkali sample 2			
Content of component 1/(mol/L)			
Average value/(mol/L)			
Content of component 2/(mol/L)			
Average value/(mol/L)			

6. Notes

(1) When the mixed alkali is composed of Na_2CO_3 and NaOH, a few more drops of phenolphthalein can be added appropriately, otherwise the measurement result of NaOH may be lower due to incomplete titration.

(2) At the first measuring point, the titration speed should be moderate, the shaking should be uniform, and excessive local HCl should be avoided to generate H_2CO_3.

7. Questions

(1) Can other indicators be selected for this experiment?

(2) Please determine the composition of the sample solution in the following five situations: (1) $V_1=V_2$, V_1, $V_2>0$; (2) $V_1=0$, $V_2>0$; (3) $V_1>0$, $V_2=0$; (4) $V_1>V_2$; (5) $V_1<V_2$.

实验十二　EDTA 标准溶液的配制与标定

一、实验目的

1. 掌握 EDTA 标准溶液的配制方法与标定方法。
2. 学会判断配位滴定的终点。
3. 了解缓冲溶液的应用。

二、实验原理

配位滴定中通常使用的配位剂是乙二胺四乙酸的二钠盐（$Na_2H_2Y·2H_2O$），其水溶液的 pH 值为 4.5 左右，如 pH 值偏低，需用 NaOH 溶液中和至 pH=5 左右，以免溶液配制后有乙二胺四乙酸析出。

配制 EDTA 标准溶液一般采用间接法。因 EDTA 标准溶液能与大多数金属离子形成 1∶1 的稳定配合物，所以可用含有金属离子的基准物如 Zn、Cu、Pb、$CaCO_3$、$MgSO_4·7H_2O$ 等，在一定 pH 值条件下，选择适当的指示剂来标定。本实验用 Zn 作基准物，选用铬黑 T(EBT) 作指示剂，在 $NH_3·H_2O$-NH_4Cl 缓冲溶液（pH=10）中进行标定，其反应如下：

滴定前（式中 In^{3-} 为金属指示剂）：$Zn^{2+} + In^{3-}$（蓝色）$= [ZnIn]^-$（酒红色）

滴定开始至终点前：$Zn^{2+} + Y^{4-} = [ZnY]^{2-}$（无色）

终点时：$[ZnIn]^-$（酒红色）$+ Y^{4-} = [ZnY]^{2-} + In^{3-}$（蓝色）

滴定终点时，溶液从酒红色变为蓝色。

三、仪器与药品

仪器：分析天平；25mL 移液管；25mL 酸式滴定管；250mL 锥形瓶；100mL 烧杯；100mL 容量瓶。

药品：纯 Zn 片；盐酸（6mol/L）；氨水（1∶1）；$NH_3 \cdot H_2O-NH_4Cl$ 缓冲溶液（pH≈10）；铬黑 T 指示剂；乙二胺四乙酸二钠盐。

四、实验步骤

1. 0.01mol/L EDTA 标准溶液的配制

称取 3.7g EDTA 二钠盐，溶于 100mL 水中，必要时可温热以加快溶解，摇匀。长期放置时，应贮存于聚乙烯瓶中。

2. 0.01mol/L Zn^{2+} 标准溶液的配制

用分析天平准确称取纯锌片 0.15～0.20g，置于 100mL 烧杯中，加入 5mL 6mol/L 盐酸，盖上表面皿，必要时水浴温热，使之完全溶解，用水洗表面皿及烧杯壁，将溶液转移于 250mL 容量瓶中，加水稀释至刻度，摇匀。计算 Zn^{2+} 标准溶液的浓度 $c_{Zn^{2+}}$。

3. EDTA 标准溶液的标定

用移液管吸取 25.00mL Zn^{2+} 标准溶液于 250mL 锥形瓶中，逐滴加入 $NH_3 \cdot H_2O$（1∶1），同时不断摇动，直至开始出现 $Zn(OH)_2$ 白色沉淀。加入 5mL $NH_3 \cdot H_2O-NH_4Cl$ 缓冲溶液（pH=10），并加 50mL 水和 3 滴铬黑 T 指示剂，用 EDTA 标准溶液滴定至溶液由酒红色变为蓝色，即为终点。平行测定三次，记下所消耗的 EDTA 标准溶液的体积（mL），计算 EDTA 标准溶液的浓度。

五、数据记录及结果处理

数据记录及结果计算见表 3-7。

表 3-7 数据记录及结果计算

试剂	1	2	3
m_{Zn}/g			
$c_{Zn^{2+}}$/(mol/L)			
V_{EDTA}/mL			
c_{EDTA}/(mol/L)			
$\bar{c}_{EDTA}$/(mol/L)			
相对平均偏差/%			

六、注意事项

1. 在配制 EDTA 溶液时要保证固体全部溶解。
2. 配位滴定要缓慢进行以保证其反应充分。

七、思考题

1. 在配位滴定中，指示剂应具备什么条件？
2. 本实验用什么方法调节 pH？

Experiment 12　Preparation and Standardization of EDTA Standard Solution

1. Purpose

(1) To master the preparation and calibration methods of EDTA standard solutions.

(2) To learn to determine the endpoint of coordination titration.

(3) To understand the application of buffer solutions.

2. Principle

The commonly used coordinating agent in coordination titration is the disodium salt of ethylenediaminetetraacetic acid ($Na_2H_2Y \cdot 2H_2O$), with an aqueous solution pH of around 4.5. If the pH value is low, it needs to be neutralized with NaOH solution to around pH 5 to prevent the precipitation of ethylenediaminetetraacetic acid after solution preparation.

Indirect method is generally used to prepare EDTA standard solution. Due to the ability of EDTA standard solution to form a 1∶1 stable complex with most metal ions, reference substances containing metal ions such as Zn, Cu, Pb, $CaCO_3$, $MgSO_4 \cdot 7H_2O$, etc. can be used to calibrate under certain pH values by selecting appropriate indicators. In this experiment, Zn was used as the reference substance and eriochrome black T(EBT) was selected as the indicator. The calibration was carried out in $NH_3 \cdot H_2O-NH_4Cl$ buffer solution (pH=10), and the reaction was as follows：

Before titration (In^{3-} is the indicator)：$Zn^{2+} + In^{3-}$ (blue) $= [ZnIn]^-$ (wine red)

From the beginning of titration to the end point：$Zn^{2+} + Y^{4-} = [ZnY]^{2-}$ (colorless)

At the end point：$[ZnIn]^-$ (wine red) $+ Y^{4-} = [ZnY]^{2-} + In^{3-}$ (blue)

At the endpoint of titration, the solution changes from wine red to blue.

3. Apparatus and Chemicals

Apparatus：Analytical balance; 25mL pipettes; 25mL acid burette; 250mL erlenmeyer flask; 100mL beaker; 100mL volumetric flask.

Chemicals：$NH_3 \cdot H_2O-NH_4Cl$ buffer solution (pH≈10); EBT indicator; pure zinc; disodium EDTA (AR); HCl (6mol/L), ammonia (1∶1).

4. Procedure

(1) Preparation of EDTA standard solution (0.01mol/L)

Weigh 3.7 g of disodium EDTA and dissolve it in 100mL of water. If necessary, warm it up to accelerate the dissolution and shake well. When placed for a long time, it should be stored in a polyethylene bottle.

(2) Preparation of Zn^{2+} standard solution (0.01mol/L)

Accurately weigh 0.15-0.20g of pure zinc flakes using an analytical balance, place them in a 100mL beaker, add 5mL of 6mol/L hydrochloric acid, cover it with a petri dish, and if necessary, warm it in the water bath to completely dissolve it. Wash the petri dish and beaker walls with water, transfer the solution to a 250mL volumetric flask, dilute with water to the mark, and shake well. Calculate the concentration of Zn^{2+} standard solution.

(3) Calibration of EDTA standard solution

Take 25.00mL of Zn^{2+} standard solution into a 250mL conical flask by a pipette, add $NH_3 \cdot H_2O$ (1:1) dropwise, and shake continuously until a white precipitate of $Zn(OH)_2$ begins to appear. Add 5mL of $NH_3 \cdot H_2O$-NH_4Cl buffer solution (pH=10), and add 50mL of water and 3 drops of eriochrome black T indicator. Titrate with EDTA standard solution until the solution changes from wine red to blue, which is the endpoint. Measure three times in parallel, record the volume (mL) of EDTA standard solution consumed, and calculate the concentration of EDTA standard solution.

5. Data Analysis

Table 3-7 lists the data record and analysis.

Table 3-7　Data record and analysis

Reagents	1	2	3
m_{Zn}/g			
$c_{Zn^{2+}}$/(mol/L)			
V_{EDTA}/mL			
c_{EDTA}/(mol/L)			
$\bar{c}_{EDTA}$/(mol/L)			
Relative average deviation/%			

6. Notes

(1) When preparing EDTA solution, ensure that all solids are dissolved.

(2) Coordination titration should be carried out slowly to ensure sufficient reaction.

7. Questions

(1) What conditions should the indicator meet in coordination titration?

(2) What method was used to adjust pH in this experiment?

实验十三 水中钙、镁含量的测定（配位滴定法）

一、实验目的

1. 了解水硬度的测定意义和常用硬度表示方法。
2. 掌握配位滴定法测定水中钙、镁含量的基本原理和方法。
3. 掌握铬黑 T 和钙指示剂的使用条件和终点变化原理。

二、实验原理

一般含有钙、镁盐类的水叫硬水，硬度可分为暂时硬度和永久硬度。暂时硬度是指钙、镁的酸式碳酸盐遇热形成碳酸盐沉淀而失去硬度；永久硬度是指钙、镁的硫酸盐、氯化物、硝酸盐等，加热也不产生沉淀。暂时硬度和永久硬度总称为"总硬度"。各国对水硬度的表示方法各有不同。其中德国硬度是较早的一种，也是我国采用较普遍的硬度之一，它以度数计，1°相当于 1L 水中含 10mg CaO 所引起的硬度。为方便计算，我国也常以 mol/L 或 mmol/L 来表示。

水中钙、镁含量可用 EDTA 法测定。测定 Ca^{2+}、Mg^{2+} 总量时，在 pH≈10 的 NH_3-NH_4Cl 缓冲溶液中，以铬黑 T 为指示剂，用 EDTA 滴定。因溶液中各配合物的稳定性存在差异（CaY^{2-}＞MgY^{2-}＞MgIn＞CaIn），铬黑 T 先与部分 Mg^{2+} 配位形成 MgIn（酒红色）。当 EDTA 滴入时，其首先与 Ca^{2+} 和 Mg^{2+} 配位，然后再夺取 MgIn 中 Mg^{2+}，使铬黑 T 游离，因此到达终点时，溶液由酒红色变为蓝色。在测定 Ca^{2+} 时，先用 NaOH 调节溶液 pH≈12～13，使 Mg^{2+} 生成难溶的 $Mg(OH)_2$ 沉淀。加入钙指示剂与 Ca^{2+} 配位溶液呈红色。滴定时，EDTA 先与游离的 Ca^{2+} 配位，然后夺取已与钙指示剂配位的 Ca^{2+}，使溶液由酒红色变成蓝色，即为滴定终点。

利用 EDTA 标准溶液用量计算钙镁总量，然后换算为相应的硬度单位。计算水硬度（°）的公式如下：

$$水硬度 = \frac{c_{EDTA} V_{EDTA} M_{CaO}}{V_{H_2O} \times 10} \times 1000$$

式中 c_{EDTA}——EDTA 标准溶液的物质的量浓度，mol/L；

V_{EDTA}——滴定时消耗 EDTA 标准溶液的体积，mL；

M_{CaO}——CaO 的摩尔质量，g/mol；

V_{H_2O}——水样体积，mL。

当 V_{EDTA} 为滴定 Ca^{2+}、Mg^{2+} 时所消耗 EDTA 标准溶液体积时，计算值为总硬度；当 V_{EDTA} 为滴定 Ca^{2+} 时所消耗 EDTA 标准溶液体积时，计算值为钙硬度。

三、仪器与药品

仪器：25mL 酸式滴定管；250mL 锥形瓶；25mL 移液管。

药品：EDTA 标准溶液；铬黑 T 指示剂；NH_3-NH_4Cl 缓冲溶液（pH=10）；NaOH（6mol/L）；钙指示剂。

四、实验步骤

1. Ca^{2+} 的测定

用移液管准确吸取 50.00mL 水样于 250mL 锥形瓶中，加入 2.00mL 6mol/L NaOH（pH=12~13），再加入 9 滴钙指示剂，溶液呈酒红色。用 EDTA 标准溶液滴定至蓝色，即为滴定终点，记录消耗 EDTA 的体积（V_{EDTA1}），平行测定三次。

2. Ca^{2+}、Mg^{2+} 总量的测定

用移液管准确吸取 50mL 水样放入 250mL 锥形瓶中，加入 5.00mL NH_3-NH_4Cl 缓冲溶液（pH=10），再加 9 滴铬黑 T 指示剂，溶液呈酒红色。用 EDTA 标准溶液滴定至蓝色，即为终点，记录消耗 EDTA 的体积（V_{EDTA2}），平行测定三次。

五、注意事项

1. EDTA 标准溶液使用前应用 Zn^{2+} 标准溶液进行标定。

2. 如溶液中存在 Fe^{3+}、Al^{3+}，可使用三乙醇胺掩蔽，但须在 pH<4 时加入，摇动后再调节 pH 至滴定酸度。Cu^{2+}、Pb^{2+}、Zn^{2+} 等重金属离子可用 KCN、Na_2S 予以掩蔽。

六、思考题

1. 如果只用铬黑 T 指示剂，能否测定 Ca^{2+} 的含量？如何测定？

2. 为什么滴定 Ca^{2+}、Mg^{2+} 总量时要控制 pH≈10，而滴定 Ca^{2+} 时，要控制 pH≈12~13？若在 pH>13 时测定 Ca^{2+}，对结果有何影响？

Experiment 13　Analysis of the Concentrations of Calcium and Magnesium Ions in a Water Sample (Complexometric Titration Method)

1. Purpose

(1) To understand the significance of measuring water hardness and commonly used hardness representation methods.

(2) To master the basic principles and methods of measuring calcium and magnesium content in water using coordination titration.

(3) To master the usage conditions and endpoint change principles of eriochrome black T and calcium indicators.

2. Principle

Generally, water containing calcium and magnesium salts is called hard water, and its hardness can be divided into temporary hardness and permanent hardness. Temporary hardness refers to the loss of hardness due to the formation of carbonate precipitates in acidic carbonates of calcium and magnesium when heated; permanent hardness refers to the sulfates, nitrides, nitrates, etc. of calcium and magnesium, which do not precipitate when heated. The temporary hardness and permanent hardness are collectively referred to as the "total hardness". The expression methods of water hardness vary among countries. Among them, German hardness is one of the earlier and more commonly used hardness in China. It is measured in degrees, and 1° is equivalent to the hardness caused by 10mg CaO in 1 L of water. For the convenience of calculation, we often use mol/L or mmol/L to represent it in China.

The calcium and magnesium content in water can be determined using EDTA method. When measuring the total amount of Ca^{2+} and Mg^{2+}, titrate with EDTA in a buffer solution of NH_3-NH_4Cl with pH≈10, using eriochrome black T as an indicator. Due to differences in the stability of various complexes in the solution (CaY^{2-}＞MgY^{2-}＞MgIn＞CaIn), eriochrome black T first coordinates with some Mg^{2+} to form MgIn (wine red). When EDTA drips in, it first coordinates with Ca^{2+} and Mg^{2+}, and then seizes Mg^{2+} from MgIn, causing eriochrome black T to dissociate. Therefore, when the endpoint is reached, the solution changes from wine red to pure blue. When determining Ca^{2+}, first adjust the solution pH≈12-13 with NaOH to generate insoluble $Mg(OH)_2$ precipitates. Add a calcium indicator to coordinate with the Ca^{2+}, let solution turn red. During titration, EDTA first coordinates with free Ca^{2+}, and then grabs the Ca^{2+} that has already coordinated with the calcium indicator, causing the solution to turn from wine red to blue, which is the titration endpoint.

The amount of EDTA standard solution was used to calculate the total amount of calcium and magnesium, and then convert it into the corresponding hardness unit. The formula for calculating water hardness (°) is as follows:

$$\text{water hardness} = \frac{c_{EDTA} V_{EDTA} M_{CaO}}{V_{H_2O} \times 10} \times 1000$$

where, c_{EDTA}—Molar concentration of EDTA standard solution, mol/L;

V_{EDTA}—Volume of EDTA standard solution used for the titration, mL;

M_{CaO}—Molar mass of CaO, g/mol;

V_{H_2O}—Volume of water sample, mL。

When V_{EDTA} is the volume of EDTA standard solution consumed for titration of Ca^{2+} and Mg^{2+}, the calculated value is the total hardness. When V_{EDTA} is the volume of EDTA standard solution consumed for titrating Ca^{2+}, the calculated value is calcium hardness.

3. Apparatus and Chemicals

Apparatus: 25mL acid burette; 25mL pipettes; 250mL erlenmeyer flask.

Chemicals: NaOH (6mol/L); NH_3-NH_4Cl buffer solution (pH≈10); EDTA standard solution; EBT indicator; calconcarboxylic acid.

4. Procedure

(1) Determination of Ca^{2+}

Pipet 50mL water sample to a 250mL erlenmeyer flask, and then add 2mL of 6mol/L

NaOH (pH = 12-13) and 9 drops of calconcarboxylic acid. Titrate the water sample with EDTA until the color of the solution turns to blue. Record the volume of EDTA (V_{EDTA}). Repeat twice.

(2) Determination of Ca^{2+} and Mg^{2+}

Pipet 50mL water sample to a 250mL erlenmeyer flask, and then add 5mL NH_3-NH_4Cl buffer solution (pH ≈ 10) and 9 drops of EBT. Titrate the water sample with EDTA until the color of the solution changes from wine red to blue. Record the volume of EDTA (V_{EDTA}). Repeat twice.

5. Notes

(1) EDTA standard solution should be calibrated by Zn^{2+} standard solution before use.

(2) If Fe^{3+} and Al^{3+} are present in the solution, triethanolamine can be used for masking, but it must be added at pH < 4, shaken, and then adjusted to the acidity for titration. Heavy metal ions such as Cu^{2+}, Pb^{2+}, Zn^{2+} can be masked by KCN and Na_2S.

6. Questions

(1) Can we determine the concentration of Ca^{2+} only with the EBT indicator? How should we do it if possible?

(2) Why should pH be controlled around 10 when titrating the total amount of Ca^{2+} and Mg^{2+}, while pH be controlled about 12-13 when titrating Ca^{2+}? If pH > 13, what is the effect to the Ca^{2+} determination?

实验十四　葡萄糖含量的测定（间接碘量法）

一、实验目的

1. 掌握间接碘量法测定葡萄糖含量的基本原理和基本方法。
2. 查阅文献资料了解其他测定葡萄糖含量的方法。

二、实验原理

I_2 与 NaOH 作用生成 NaIO（次碘酸钠）：

$$I_2 + 2NaOH = NaIO + NaI + H_2O$$

而 $C_6H_{12}O_6$（葡萄糖）能定量地被 NaIO 氧化：

$$C_6H_{12}O_6 + NaIO = C_6H_{12}O_7 + NaI$$

未与 $C_6H_{12}O_6$ 反应剩余的 NaIO 发生歧化反应生成 $NaIO_3$ 和 NaI，在酸性条件下，继续转化成 I_2 析出：

$$3NaIO = NaIO_3 + 2NaI（歧化反应）$$

$$NaIO_3 + 5NaI + 6HCl = 3I_2 + 6NaCl + 3H_2O$$

因此，只要用 $Na_2S_2O_3$ 标准溶液滴定析出的 I_2，便可计算出剩余的 NaIO，进而推算出 $C_6H_{12}O_6$ 的含量。反应方程式如下：

$$I_2 + 2Na_2S_2O_3 =\!\!=\!\!= Na_2S_4O_6 + 2NaI$$

在上述一系列反应中，1mol I_2 产生 1mol NaIO，而 1mol NaIO 与 1mol $C_6H_{12}O_6$ 作用。因此，1mol $C_6H_{12}O_6$ 与 1mol I_2 相当。加入 I_2 的总量减去与 $Na_2S_2O_3$ 反应的 I_2，即为与 $C_6H_{12}O_6$ 反应的 I_2 量，按下式计算样品中葡萄糖的含量（g/L）：

$$\rho_{C_6H_{12}O_6} = \frac{(c_{I_2}V_{I_2} - \frac{1}{2}c_{Na_2S_2O_3}V_{Na_2S_2O_3})M_{C_6H_{12}O_6}}{25.00}$$

三、仪器与药品

仪器：250mL 碘量瓶；25mL 移液管；25mL 酸式滴定管。

药品：HCl（6mol/L）；NaOH（2mol/L）；$Na_2S_2O_3$ 标准溶液（0.1mol/L）；I_2 溶液（0.05mol/L）；淀粉溶液（0.5%）；葡萄糖（w 为 0.50）。

四、实验步骤

1. 葡萄糖溶液的配制

计算配制 250mL w 为 0.005 的葡萄糖溶液所需 w 为 0.50 葡萄糖溶液的体积。量取计算体积的葡萄糖溶液加入 250mL 容量瓶中，加入去离子水稀释至刻度，摇匀。

2. 葡萄糖溶液含量的测定

移取 25.00mL 待测液于碘量瓶中，准确加入 I_2 标准溶液 25.00mL，一边摇动一边慢慢滴加 2mol/L NaOH 溶液（加碱的速度不能过快，否则过量的 NaIO 来不及氧化 $C_6H_{12}O_6$，使测定结果偏低），直至溶液呈淡黄色。将碘量瓶盖好暗处放置 10min 后，加入 2.00mL 6mol/L HCl 使溶液呈酸性，随后用 $Na_2S_2O_3$ 溶液滴定，至溶液呈浅黄色，加入 2.00mL 淀粉指示剂，继续滴至蓝色消失，振荡 30s 溶液仍为无色，即为滴定终点，记下消耗 $Na_2S_2O_3$ 的体积，平行测定三次。

五、数据记录及结果处理

数据记录与结果计算见表 3-8。

表 3-8　数据记录与结果计算

试剂	1	2	3
$V_{Na_2S_2O_3}$/mL			
$\rho_{C_6H_{12}O_6}$/(g/L)			
$\bar{\rho}_{C_6H_{12}O_6}$/(g/L)			
相对平均偏差/%			

六、注意事项

1. 淀粉指示剂应在临近终点时加入。
2. 滴定需要在中性或弱酸性溶液中进行。

七、思考题

1. 碘溶液为什么要被保存在带玻璃塞的棕色瓶中？
2. 标定碘溶液时，既可以用硫代硫酸钠滴定碘溶液，也可以用碘溶液滴定硫代硫酸钠溶液，且都用淀粉做指示剂。但在两种情况下加入淀粉指示剂的时间是否相同？

Experiment 14   Determination of Glucose (Indirect Iodometry)

1. Purpose

(1) To understand the procedure and principle of determining glucose.
(2) To master the other five methods of determining glucose.

2. Principle

I_2 reacts with NaOH to generate NaIO (sodium hypoiodite):

$$I_2 + 2NaOH = NaIO + NaI + H_2O$$

And $C_6H_{12}O_6$ (glucose) can be quantitatively oxidized by NaIO:

$$C_6H_{12}O_6 + NaIO = C_6H_{12}O_7 + NaI$$

The remaining NaIO that did not react with $C_6H_{12}O_6$ undergoes disproportionation reaction to generate $NaIO_3$ and NaI. Under acidic conditions, it continues to transform into I_2 and precipitate:

$$3NaIO = NaIO_3 + 2NaI \text{(Disproportionation reaction)}$$
$$NaIO_3 + 5NaI + 6HCl = 3I_2 + 6NaCl + 3H_2O$$

Therefore, as long as the precipitated I_2 is titrated with $Na_2S_2O_3$ standard solution, the remaining NaIO can be calculated, and then the content of $C_6H_{12}O_6$ can be calculated. The reaction equation is as follows:

$$I_2 + 2Na_2S_2O_3 = Na_2S_4O_6 + 2NaI$$

In the above series of reactions, 1 mol I_2 produces 1 mol NaIO, which interacts with 1 mol $C_6H_{12}O_6$. Therefore, 1 mol $C_6H_{12}O_6$ is equivalent to 1 mol I_2. The total amount of I_2 added minus the amount of I_2 reacting with $C_6H_{12}O_6$ is the amount of I_2 reacting with $C_6H_{12}O_6$. Calculate the glucose content in the sample (g/L) using the following formula:

$$\rho_{C_6H_{12}O_6} = \frac{(c_{I_2}V_{I_2} - \frac{1}{2}c_{Na_2S_2O_3}V_{Na_2S_2O_3})M_{C_6H_{12}O_6}}{25.00}$$

3. Apparatus and Chemicals

Apparatus: 25mL pipette; 250mL iodine flask; 25mL acid burette.

Chemicals: HCl (6mol/L); NaOH (2mol/L); $Na_2S_2O_3$ standard solution (0.1mol/L); Iodine (0.05mol/L); starch indicator (0.5%); glucose solution ($w = 0.50$).

4. Procedure

(1) Preparation of glucose solution

Calculate the required volume of glucose solution ($w=0.50$) to prepare 250mL of glucose solution ($w=0.005$). Measure the volume of glucose solution and add it to a 250mL volumetric flask. Dilute with deionized water to the mark and shake well.

(2) Determination of glucose solution content

Transfer 25.00mL of the test solution into an iodine volumetric flask, accurately add 25.00mL of I_2 standard solution, and slowly add 2mol/L NaOH solution while shaking (the speed of adding alkali should not be too fast, otherwise excessive NaIO will not be able to oxidize $C_6H_{12}O_6$ in time, resulting in low measurement results), until the solution turns pale yellow. After capping the iodine volumetric flask and leaving it in the dark for 10minutes, add 2.00mL of 6mol/L HCl to make the solution acidic. Then titrate with $Na_2S_2O_3$ solution until the solution turns light yellow. Add 2.00mL of starch indicator and continue to drip until the blue color disappears. Shake for 30 seconds if the solution remains colorless, which is the titration endpoint. Record the volume of $Na_2S_2O_3$ consumed and measure three times in parallel.

5. Data Analysis

Table 3-8 lists the data record and analysis.

Table 3-8 Data record and analysis

Reagents	1	2	3
$V_{Na_2S_2O_3}$/mL			
$\rho_{C_6H_{12}O_6}$/(g/L)			
$\overline{\rho}_{C_6H_{12}O_6}$/(g/L)			
Relative average deviation/%			

6. Notes

(1) Starch indicator should be added near the endpoint.

(2) Titration needs to be performed in neutral or weakly acidic solutions.

7. Questions

(1) Why is iodine solution stored in a brown bottle with a glass stopper?

(2) In calibration of iodine solution, sodium thiosulfate and iodine solution can be used to titrate each other, in both cases starch can be used as an indicator. But is the time for adding starch indicator the same in both cases?

实验十五　化学需氧量（COD）的测定（高锰酸钾法）

一、实验目的

1. 掌握酸性高锰酸钾法测定水中化学需氧量（COD）的原理和方法。
2. 了解测定 COD 的意义。

二、实验原理

化学需氧量是指用适当氧化剂处理水样时，水样中需氧污染物所消耗的氧化剂的量，通常以相应的氧量（单位为 mg/L）来表示。COD 是表示水体或污水污染程度的重要综合性指标之一，是环境保护和水质控制中经常需要测定的项目。COD 值越高，说明水体污染越严重。COD 的测定分为酸性高锰酸钾法、碱性高锰酸钾法和重铬酸钾法。$KMnO_4$ 法得到的值记为高锰酸钾指数，仅适用于污染不太重的地表水、饮用水和生活污水等。

本实验采用酸性高锰酸钾法。即在酸性条件下，向被测水样中定量加入高锰酸钾溶液，加热水样，使高锰酸钾与水样中有机污染物充分反应，过量的高锰酸钾用一定量的草酸钠还原，最后用高锰酸钾溶液返滴过量的草酸钠，由此计算出水样的耗氧量。反应方程式为：

$$2MnO_4^- + 5C_2O_4^{2-} + 16H^+ = 2Mn^{2+} + 10CO_2\uparrow + 8H_2O$$

三、仪器与药品

仪器：25mL 酸式滴定管；250mL 锥形瓶；100mL 量筒；25mL 移液管；水浴锅。

药品：$KMnO_4$ 溶液（0.0050mol/L）；硫酸（1∶2）；硝酸银溶液（w 为 0.10）；草酸钠标准溶液（0.0130mol/L）。

四、实验步骤

1. 取 50.00mL 水样加入 250mL 锥形瓶中，用去离子水稀释至 100mL，加入硫酸（1∶2）10.00mL，再加入 w 为 0.10 的硝酸银溶液 5.00mL，摇匀后准确加入 0.0050mol/L $KMnO_4$ 溶液 10.00mL（V_1）。将锥形瓶置于沸水浴中加热 30min，充分氧化需氧污染物。稍冷后（~80℃），加入 0.0130mol/L $Na_2C_2O_4$ 标准溶液 10.00mL，摇匀（此时溶液应为无色），趁热（此时温度不应低于 70℃，否则需加热）用 0.0050mol/L $KMnO_4$ 溶液滴定至微红色，30s 内不褪色即为滴定终点，记下 $KMnO_4$ 溶液的用量为 V_2。

2. 在 250mL 锥形瓶中加入去离子水 50.00mL 和硫酸（1∶2）10.00mL，移入 0.0130mol/L $Na_2C_2O_4$ 标准溶液 10.00mL，摇匀，在 70~80℃ 的水浴中，用 0.0050mol/L $KMnO_4$ 溶液滴定至溶液呈微红色，30s 内不褪色即为滴定终点，记下 $KMnO_4$ 溶液的用量为 V_3。

3. 在 250mL 锥形瓶中加入去离子水 50.00mL 和硫酸（1∶2）10.00mL，在 70~80℃ 的水浴中，用 0.0050mol/L $KMnO_4$ 溶液滴定至溶液微红色，30s 内不褪色即为终点，记下 $KMnO_4$ 溶液的用量为 V_4。按下式计算化学需氧量 $COD_{(Mn)}$：

$$COD_{Mn} = \frac{[(V_1+V_2-V_4)\cdot f-10.00]\times c_{Na_2C_2O_4}\times 16.00\times 1000}{V_s}$$

式中，$f=\dfrac{10}{V_3-V_4}$，即 1mL $KMnO_4$ 相当于 fmL $Na_2C_2O_4$ 标准溶液；V_1+V_2 是与 50mL 水样、50mL 去离子水以及 10mL 草酸钠反应的高锰酸钾溶液体积；V_3 为与 50mL 去离子水和 10mL 草酸钠反应的高锰酸钾溶液体积；V_4 为与 50mL 去离子水反应消耗的高锰酸钾溶液体积；V_s 为水样体积（50mL）；16.00 是 $\dfrac{1}{2}O_2$ 的摩尔质量。

五、注意事项

1. 在水浴加热完毕后，溶液仍应保持淡红色，如变浅或全部褪去，说明高锰酸钾的用量不够。此时，应将水样稀释倍数加大后再测定。

2. 水样中 Cl^- 在酸性高锰酸钾中被氧化，使结果偏高。

六、思考题

1. 测定水样的耗氧量时，是否一定要加入硝酸银？加入硝酸银的作用是什么？
2. 哪些因素影响 COD 的测定结果？为什么？

Experiment 15　Determination of Chemical Oxygen Demand（COD）（$KMnO_4$ Titrimetry）

1. Purpose

（1）To master the principle and method of measuring chemical oxygen demand（COD）in water using acidic potassium permanganate method.

（2）To understand the significance of measuring chemical oxygen demand（COD）.

2. Principle

Chemical oxygen demand（COD）refers to the amount of oxidant consumed by aerobic pollutants in water samples when treated with appropriate oxidants，usually expressed in the corresponding oxygen content（mg/L）. COD is one of the important comprehensive indicators that represent the degree of water or sewage pollution，and is often measured in environmental protection and water quality control. The higher the COD value，the more severe the water pollution. The determination of COD can be divided into acidic potassium permanganate method，alkaline potassium permanganate method，and potassium dichromate method. The value obtained by the $KMnO_4$ method is denoted as the potassium permanganate index，which is only applicable to surface water，drinking water，and domestic sewage that are not heavily polluted.

This experiment adopts the acidic potassium permanganate method. Under acidic conditions，a potassium permanganate solution is quantitatively added to the tested water sample，and the water sample is heated to fully react with organic pollutants in the water sample. The excess potassium permanganate is reduced with a certain amount of sodium oxalate，and finally，the excess sodium oxalate is dripped back with the potassium permanganate solution to calculate the oxygen consumption of the water sample. The reaction equation is as follows:

$$2MnO_4^- + 5C_2O_4^{2-} + 16H^+ = 2Mn^{2+} + 10CO_2 + 8H_2O$$

3. Apparatus and Chemicals

Apparatus: 25mL acid burette, 250mL erlenmeyer flask, 100mL measuring cylinder, 25mL pipet, water bath.

Chemicals: 0.0050mol/L $KMnO_4$, H_2SO_4 (1:2), $AgNO_3$ ($w=0.10$), $Na_2C_2O_4$ standard solution (0.0130mol/L).

4. Procedure

(1) Add 50.00mL of water sample to a 250mL conical flask. Dilute to 100mL with deionized water, add 10.00mL of sulfuric acid (1:2) and 5.00mL of silver nitrate solution ($w=0.10$). Shake well and accurately add 10.00mL (V_1) of 0.0050mol/L $KMnO_4$ solution. Heat the conical flask in a boiling water bath for 30minutes to fully oxidize the aerobic pollutants. After slightly cooling (~80°C), add 10.00ml of 0.01300mol/L $Na_2C_2O_4$ standard solution, shake well (the solution should be colorless at this time), and titrate with 0.0050mol/L $KMnO_4$ solution while it is hot (the temperature should not be lower than 70°C, otherwise heating is required) until it turns slightly red. If it does not turn red within 30 seconds, it is the titration endpoint. Record the amount of $KMnO_4$ solution as V_2.

(2) Add 50mL distilled water and 10mL of H_2SO_4 (1:2) to a 250mL conical flask, then transfer 10.00mL 0.0130mol/L $Na_2C_2O_4$ standard solution to it, shake well and titrate with 0.005mol/L $KMnO_4$ solution at 70-80°C until the color turns slightly red. The titration reaches end point when there is no color within 30 seconds. Record the amount of $KMnO_4$ solution as V_3.

(3) Add 50.00mL of deionized water and 10.00mL of sulfuric acid (1:2) to a 250mL conical flask, and titrate with 0.0050mol/L $KMnO_4$ solution in a water bath at 70-80°C until the solution turns slightly red. The end point is arrived if the solution not fading within 30 seconds. Record the amount of $KMnO_4$ solution used as V_4.

Calculate the chemical oxygen demand COD (Mn) using the following formula:

$$COD_{Mn} = \frac{[(V_1+V_2-V_4) \cdot f - 10.00] \times c_{Na_2C_2O_4} \times 16.00 \times 1000}{V_s}$$

Where, $f = \dfrac{10}{V_3-V_4}$, 1mL of $KMnO_4$ is equivalent to f mL of $Na_2C_2O_4$ standard solution. V_1+V_2 is the volume of potassium permanganate solution reacted with 50mL of water sample, 50mL of deionized water, and 10mL of sodium oxalate. V_3 is the volume of potassium permanganate solution reacted with 50mL of deionized water and 10mL of sodium oxalate. V_4 is the volume of potassium permanganate solution consumed by reaction with 50mL of deionized water. V_s is the volume of the water sample (50mL). 16.00 are the Molar mass of $\frac{1}{2}O_2$.

5. Notes

(1) After the water bath is heated, the solution should still remain light red. If it becomes lighter or completely fades, it indicates that the amount of potassium permanganate is not enough. At this point, the dilution ratio of the water sample should be increased before measurement.

(2) Cl^- in the water sample is easy to be oxidized in acidic potassium permanganate, resulting in higher results.

6. Questions

(1) Is it necessary to add silver nitrate when measuring the oxygen consumption of water samples? What is the effect of adding silver nitrate?

(2) What factors affect the chemical oxygen demand (COD) measurement results? Why?

第四章
综合拓展实验

实验十六 纳米硒的制备与表征

一、实验目的

1. 了解纳米硒的性质。
2. 掌握生物法制备纳米硒的基本原理和方法。

二、实验原理

硒（Se）是一种生命不可或缺的微量非金属元素，也是自然界重要的元素之一。硒（Se）通常以无机和有机两种形态存在于自然界中，无机形态包含 Se(-2)、Se(0)、Se(+4) 和 Se(+6) 四种价态。相比于其他价态，单质纳米硒（Selenium nanoparticles，SeNPs）具有易吸收、毒性低的特性，其在抗氧化、抗肿瘤和抗微生物感染等领域有着巨大的发展潜力。目前，SeNPs 的合成方法主要有物理法、化学法和生物法。物理法制备 SeNPs 的操作过程相对简便，但对设备要求较高；化学法工艺较为繁琐，且在制备过程中容易产生有毒物质，对环境及人体危害较大；生物法因具有反应条件温和，环境友好，稳定性强等优点，逐渐成为制备 SeNPs 的新趋势。生物法包括微生物法和植物法，但微生物法合成纳米硒条件较苛刻，需要严格控制 pH、温度等培养条件，植物法合成纳米硒避免了微生物培养的复杂过程，且所得纳米硒具有典型的生物合成产物特性，例如生物相容性、抗菌活性等。因此，植物分子介导的纳米硒生物合成更具吸引力。本实验以柠檬提取液为原料，利用其中丰富的多酚、黄酮、VC 等还原性物质与亚硒酸反应制备纳米硒。

三、主要仪器及试剂

仪器：紫外可见分光光度计、透射电镜、恒温水浴箱、平板磁力搅拌器、超声清洗机、离心机、50mL 和 250mL 烧杯、10mL 移液管、0.45μm 微滤膜。

试剂：柠檬，市售；Na_2SeO_3（50mmol/L）；氨水（质量分数 25%）。

四、实验步骤

1. 柠檬提取液的制备

柠檬洗净去皮，称量10g柠檬果肉，置于烧杯中，加入50mL去离子水，料液比（质量：体积）为1:5，在60℃的水浴锅中浸提30min，在8000r/min下离心10min后，清液用孔径0.45μm微滤膜过滤备用。

2. 纳米硒的制备

将10mL柠檬提取液与10mL 50mmol/L的亚硒酸溶液混合，超声10min后，室温下磁力搅拌50min，观察反应混合物颜色变化，利用UV-Vis跟踪纳米硒的形成，反应结束后，对反应混合液进行离心，分离上清液，将得到的纳米硒粒子用纯净水清洗3遍，最后冷冻干燥48h，即得红色纳米硒粒子。

3. pH对纳米硒合成的影响

纳米硒制备方法如2所述，反应时间1h。不同的是取3份10mL柠檬提取液（料液比为1:5）分别与10mL的亚硒酸溶液（50mmol/L）在室温下混合，一份不加氨水，另外2份用氨水分别调节pH为6和10，反应过程取样观察溶液颜色变化，并借助UV-Vis吸收峰位置和吸收强度确定最合适pH。

4. 纳米硒形貌表征

利用透射电镜检测纳米硒的粒径和形貌。

五、思考题

1. 除柠檬提取物外，还可以利用什么植物提取物制备纳米硒？
2. 植物提取物还原纳米硒的机理是什么？

实验十七　上转换纳米粒子的制备与表征

一、实验目的

1. 了解上转换纳米发光材料的发光特点。
2. 了解水热法制备上转换纳米粒子的原理及方法。

二、实验原理

上转换发光材料就是指材料受到光激发时，能基于双光子或多光子机制，把长波辐射转换成短波辐射，即材料受到光激发时，可以发射出比激发波长短的荧光，其本质是一种反斯托克斯（Anti-Stokes）发光。这种材料具有可见光区发光、无散射光、背景干扰小和信噪比高等优点，被广泛应用于防伪、温度传感、成像治疗以及生物分析等领域。迄今为止，人们提出了上转换纳米材料的多种化学合成方法，如热分解法、高温共沉淀法、水热合成法、溶胶凝胶法、阳离子交换法和离子液体合成法等。不同的合成方法会影响制备的上转换纳米材料的发光效率和表面性能。此外，不同的稀土元素掺杂，以及不同比例的稀土元素掺杂均会影响制备的上转换纳米材料的发光颜色。本实验选择方便简单的水热合成法，通过Yb/Tm和Er/Tm共掺杂，实现蓝光发射和黄光发射两种荧光颜色的上转换纳米材料的制备。

三、主要仪器及试剂

仪器：移液枪；烧杯；烘箱；离心机；高温反应釜；紫外透射分析仪；荧光光谱仪；离心管。

试剂：Y(NO$_3$)$_3$ 溶液（0.2mol/L）；YbCl$_3$ 溶液（0.2mol/L）；ErCl$_3$ 溶液（0.2mol/L）；Tm(NO$_3$)$_3$ 溶液（0.1mol/L）；柠檬酸三钠溶液（0.1mol/L）；十六烷基三甲基溴化铵（CTAB）；无水乙醇。

四、实验步骤

1. 蓝色荧光上转换纳米粒子的制备

分别将 2.1mL 超纯水、1.2mL Y(NO$_3$)$_3$ 溶液（0.2mol/L）、1mL YbCl$_3$ 溶液（0.2mol/L）、100μL Tm(NO$_3$)$_3$ 溶液（0.1mol/L）和 1.75mL 柠檬酸三钠溶液（0.1mol/L）依次加入 50mL 烧杯中，混匀。随后向其中依次加入 0.1g CTAB 以及 15mL 无水乙醇，混匀后逐滴加入 6mL NaF 溶液（1mol/L）并持续搅拌 2h。搅拌结束后滴加 1mL 浓硝酸，即得到蓝色荧光上转换纳米粒子的前驱体。

将上述制备出的前驱体转移至 50mL 高温反应釜中，放入 200℃ 烘箱中反应 2h。冷却至室温后收集产物并于 12000r/min 转速下离心 25min，然后依次用超纯水及无水乙醇洗涤沉淀 1~2 次，即得到蓝色荧光的 NaYF$_4$：Yb/Tm 上转换纳米粒子。材料经 XRD 检测确定其结构及纯度，于 980nm 激光器照射下观察其荧光颜色并在荧光光谱仪上测定发射光谱。

2. 黄色荧光上转换纳米粒子的制备

按照 Er：Tm＝99.5：0.5 的比例重复上述步骤，即制备出黄色荧光的 NaYF$_4$：Er/Tm 上转换纳米粒子。详细过程：分别将 2.1mL 超纯水、1.2mL Y(NO$_3$)$_3$ 溶液（0.2mol/L）、1mL ErCl$_3$ 溶液（0.2mol/L）、10μL Tm(NO$_3$)$_3$ 溶液（0.1mol/L）和 1.75mL 柠檬酸三钠溶液（0.1mol/L）依次加入 50mL 烧杯中，混匀。随后向其中依次加入 0.1g CTAB 以及 15mL 无水乙醇，混匀后逐滴加入 6mL NaF 溶液（1mol/L）并持续搅拌 2h。搅拌结束后滴加 1mL 浓硝酸，即得到黄色荧光上转换纳米粒子的前驱体。

将上述制备出的前驱体转移至 50mL 高温反应釜中，放入 200℃ 烘箱中反应 2h。冷却至室温后收集产物并于 12000r/min 转速下离心 25min，然后依次用超纯水及无水乙醇洗涤沉淀 1~2 次，即得到黄色荧光的 NaYF$_4$：Er/Tm 上转换纳米粒子。收集材料于 5.0mL 离心管中并置于紫外透射分析仪下观察其荧光颜色。

五、思考题

1. 上转换发光材料有什么特点？
2. 多种稀土元素的掺杂能否实现白光上转换纳米粒子的制备？

实验十八　金属酞菁的制备与表征

一、实验目的

1. 通过合成金属酞菁配合物，掌握大环配合物的模板合成方法。

2. 了解金属模板反应在无机合成中的应用。
3. 进一步熟练掌握无机合成中的常规操作方法和技能。

二、实验原理

酞菁类化合物（MPc）是一类重要的四氮大环配体，具有高度共轭π体系。它能与金属离子形成金属酞菁配合物，其分子结构式如图4-1。金属酞菁是近年来广泛研究的经典金属类大环配合物中的一类，其基本结构和天然金属卟啉相似，具有良好的热稳定性，因此在光电转换、催化活性小分子、信息存储、生物模拟及工业染料等方面有着重要的应用。金属酞菁的合成方法主要是模板法，即通过简单配体单元与中心金属离子配位作用，然后再结合成金属大环配合物，金属离子起到模板的作用。

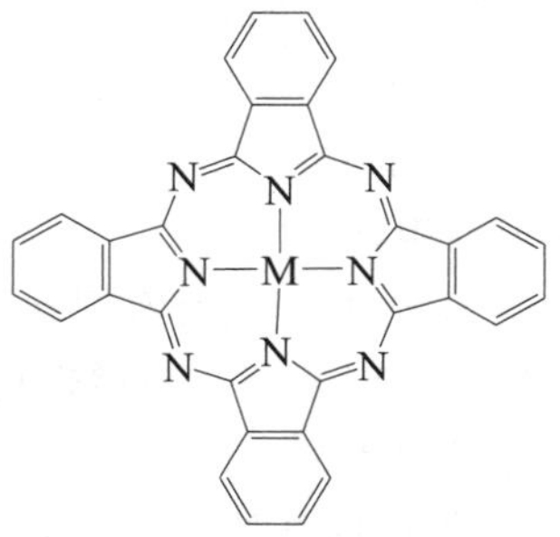

图4-1　金属酞菁分子结构

合成反应途径：

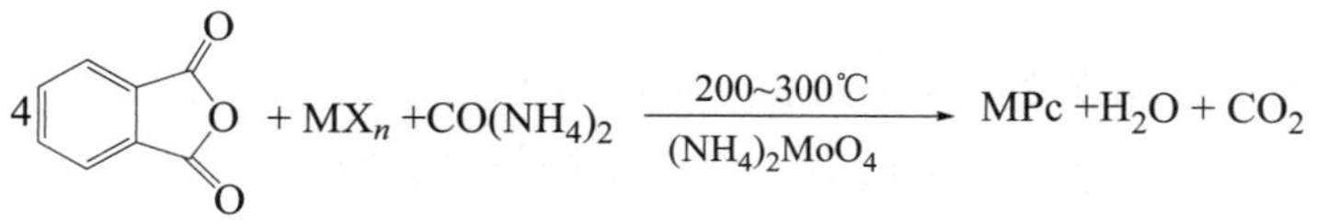

$$4 \text{(邻苯二甲酸酐)} + MX_n + CO(NH_2)_2 \xrightarrow[(NH_4)_2MoO_4]{200\sim300℃} MPc + H_2O + CO_2$$

本实验以邻苯二甲酸酐、尿素、无水氯化钴为原料，以钼酸铵为催化剂，采用模板法合成酞菁钴。用浓硫酸再沉淀法提纯产物，通过红外光谱、紫外-可见光谱对产物进行表征。

三、仪器与药品

仪器：分析天平；可控温电热套；研钵；250mL 三口烧瓶；布氏漏斗；抽滤瓶；冷凝管；圆底烧瓶；表面皿；铁架台；真空干燥箱；循环水真空泵；紫外-可见分光光度计；红外光谱仪。

药品：邻苯二甲酸酐（分析纯）；无水乙醇（分析纯）；钼酸铵（分析纯）；无水氯化钴（分析纯）；尿素；煤油；HCl（2%）。

四、实验步骤

1. 酞菁钴粗产品的制备

称取邻苯二甲酸酐 5g，尿素 9g 及钼酸铵 0.4g 于研钵中研细后加入 0.8g 无水氯化钴，混匀后马上移入 250mL 三口烧瓶中，加入 70mL 煤油，加热（200℃）回流 2h 左右，在溶液由蓝色变为紫红色后停止加热，冷却至 70℃，加入 10~15mL 无水乙醇稀释后趁热抽滤，并用乙醇洗涤 2 次，得到粗产品。

2. 粗产品提纯

将滤饼加入 2% 盐酸煮沸后趁热抽滤，再将滤饼加入去离子水中煮沸后抽滤，最后加入碱液中煮沸抽滤，重复上述步骤 2~3 次，直至滤液接近无色且 pH 呈中性。

将产品放在表面皿上 70℃ 真空干燥 6h，称重并计算产率。

3. 样品的表征与分析

取少量样品与干燥 KBr 混合研磨并压片，作红外光谱分析。取少量样品溶于二甲基亚砜中，作紫外可见光谱分析。

五、数据记录及结果处理

1. 红外光谱

金属酞菁特征吸收带主要分布在 4 个区域：

① 在 3030cm^{-1} 处的一组峰是芳环上的 C—H，谱带较尖锐。

② 在 1580cm^{-1} 和 1600cm^{-1} 处各有一吸收峰，这是由芳香环上 C═C 以及 C═N 的伸缩振动引起。

③ 在低频区可看到在与金属酞菁相应的位置上，自由酞菁的谱图上是两个对应的谱带，相比于金属酞菁，该谱带更偏于较高频率，不同中心金属使金属酞菁吸收峰向高频发生移动的程度也不同。

④ 在远红外区，骨架振动吸收带主要出现在 150～200cm^{-1} 区间，对于 Fe、Co、Ni 和 Cu 金属酞菁，这组谱带为金属-配体-配体振动，自由酞菁不出现该谱带。金属酞菁中的金属-配体-配体的振动频率按下列顺序向高频方向发生移动：Zn＞Pd＞Pt＞Cu＞Fe＞Co＞Ni。

2. 紫外光谱

一般金属酞菁的 B 带在 250～300nm，而 Q 带约在 700～800nm。B 带受中心金属以及酞菁环的变化如取代、加氢等影响较小，而 Q 带则较易受影响。

六、注意事项

1. 加入无水氯化钴迅速混匀并装入提前干燥好的三口烧瓶中，马上加入煤油，以防止吸收空气中的水分。

2. 回流一定要等到溶液由蓝色变为紫红色后再停止。

3. 重复抽滤要等到滤液颜色接近无色时再停止，否则杂质太多，影响随后的表征。

七、思考题

1. 从酞菁钴的紫外可见光谱可以得出哪些信息？

2. 如何处理实验过程中产生的废液？不经处理的废液直接倒入水槽会造成哪些危害？

实验十九　葡萄糖酸锌的制备与含量测定

一、实验目的

1. 了解葡萄糖酸锌的制备方法。
2. 掌握锌盐含量的测定方法。

二、实验原理

本实验采用葡萄糖酸钙与硫酸锌直接反应制取葡萄糖酸锌。葡萄糖酸钙与等物质的量硫酸锌反应，生成葡萄糖酸锌和硫酸钙沉淀。其反应式如下：

$$Ca(C_6H_{11}O_7)_2 + ZnSO_4 \rightleftharpoons Zn(C_6H_{11}O_7)_2 + CaSO_4 \downarrow$$

分离硫酸钙沉淀后，得到葡萄糖酸锌。

采用配位滴定的方法测定产品中锌含量，用 EDTA 标准溶液在 NH_3-NH_4Cl 弱碱性条件下滴定葡萄糖酸锌，根据所消耗滴定剂 EDTA 的量计算锌含量。

按下式计算样品中锌含量（%）：

$$w_{Zn} = \frac{c_{EDTA} V_{EDTA} M_{Zn} \times 4}{m_{葡萄糖酸锌} \times 1000} \times 100\%$$

三、仪器与药品

仪器：分析天平；水浴锅；布氏漏斗；抽滤瓶；蒸发皿；温度计；100mL 量筒；100mL 容量瓶；25mL 移液管；250mL 锥形瓶；100mL 烧杯；25mL 酸式滴定管。

药品：$ZnSO_4 \cdot 7H_2O$（分析纯），乙醇（95%），EDTA 标准溶液（0.05mol/L），NH_3-NH_4Cl 缓冲溶液（pH=10），铬黑 T 指示剂。

四、实验步骤

1. 葡萄糖酸锌的制备

准确称取 13.4g $ZnSO_4 \cdot 7H_2O$ 置于烧杯中，加入 80mL 去离子水，80～90℃水浴加热至完全溶解。将烧杯放入 90℃恒温水浴中，逐渐加入 20g 葡萄糖酸钙，并不断搅拌。90℃水浴保温 20min 后趁热抽滤。滤液移至蒸发皿中并在沸水浴上浓缩至黏稠状。冷至室温，加 95%乙醇 20mL 并不断搅拌，此时有大量胶状葡萄糖酸锌析出。充分搅拌后，用倾析法去除乙醇液。再在沉淀上加 95%乙醇 20mL，充分搅拌后，沉淀慢慢转变成晶体状，抽干，即得粗品（母液回收）。再将粗品加水 20mL，加热至溶解，趁热抽滤，滤液冷至室温，加入 20mL 95%乙醇，充分搅拌，结晶析出后，抽干，50℃烘干。

2. 锌含量的测定

准确称取 1.6000g 葡萄糖酸锌，置于小烧杯中，加水溶解后，转入 100mL 容量瓶中，以去离子水稀释至标线，摇匀。准确移取 25.00mL 溶液于 250mL 锥形瓶中，加入 10mL NH_3-NH_4Cl 缓冲溶液，4 滴铬黑 T 指示剂，然后用 0.05mol/L EDTA 标准溶液滴定，滴至溶液由红色刚好转变为蓝色为止，根据所用 EDTA 标准溶液体积计算样品中锌含量。平行测定三次。

五、数据记录及结果处理

锌含量的测定数据见表 4-1。

表 4-1 锌含量的测定数据

试剂	1	2	3
$V_{葡萄糖酸锌}$/mL			
V_{EDTA}/mL			
w_{Zn}/%			
$\overline{w}_{Zn}$/%			
相对平均偏差/%			

六、注意事项

反应需在90℃恒温水浴中进行,温度过高葡萄糖酸锌会分解;过低则反应速度较慢。

七、思考题

1. 在沉淀与结晶葡萄糖酸锌时,加入95%乙醇的作用是什么?
2. 在葡萄糖酸锌的制备中,为什么必须在热水浴中进行?

实验二十　石墨相氮化碳量子点的制备与水质硫化物的测定

一、实验目的

1. 了解水热法制备石墨相氮化碳量子点的原理及方法。
2. 了解荧光法测定水质硫化物的原理及方法。
3. 熟悉荧光分光光度计的使用方法。

二、实验原理

水质硫化物污染作为生活中分布最广泛的污染之一,包括溶解性的硫化氢、酸溶性的金属硫化物、不溶性的硫化物和有机硫化物。通常需要进行测定的硫化物是指溶解性的和酸溶性的硫化物。生产生活中检测水质硫化物的方法多种多样,主要有分光光度法、离子色谱法、气相分子吸收法等,本实验将利用硫化物与汞离子(Hg^{2+})之间的特异性间接检测。硫化物可以与汞离子(Hg^{2+})结合产生HgS沉淀,其中Hg^{2+}可以使石墨相氮化碳量子点的荧光发生猝灭。因此,固定体系中Hg^{2+}的浓度,随着加入的硫化物浓度的增大,与硫化物结合的Hg^{2+}增多,而作用于石墨相氮化碳量子点上的Hg^{2+}就越少,因此石墨相氮化碳量子点由于Hg^{2+}导致的荧光猝灭现象会逐渐恢复,即材料的荧光逐渐增强。基于此,可以实现对硫化物的定量检测。在一定的浓度范围内,量子点的荧光强度与硫化物含量成正比,这是本实验检测硫化物的基础。

三、主要仪器及试剂

仪器:荧光分光光度计;分析天平;烘箱;离心机;高温反应釜;1mL、5mL、10mL移液管;25mL容量瓶。

试剂:尿素;柠檬酸钠;氯化汞溶液(2μmol/L);硫化钠溶液(5μmol/L)。

四、实验步骤

1. 石墨相氮化碳量子点的制备

称取1.01g尿素和0.81g柠檬酸钠,于玛瑙研钵中充分研磨混合。将混合物转移至50mL高温反应釜中,放入180℃烘箱中反应1h。冷却至室温后收集产物并于8000r/min转速下离心10min,得到的黄色上清液即为石墨相氮化碳量子点。

2. 工作曲线的制作

分别准确移取初始硫化钠溶液1.00mL、2.50mL、5.00mL、10.00mL于4个25mL容

量瓶中,用去离子水稀释至刻度,摇匀。连同未稀释的硫化钠溶液,可得到 5 种不同浓度的溶液,由稀到浓依次编号为 1、2、3、4、5。取上述 1~5 号溶液 5.00mL 与 5.00mL 的氯化汞溶液（2μmol/L）混合,静置 3min,再加入 1.00mL 的石墨相氮化碳量子点,混匀。用荧光分光光度计,在激发波长 430nm、厚度为 1cm 的比色皿条件下,测定一系列混合物溶液的荧光强度。以硫化物的浓度为横坐标,相应的荧光强度为纵坐标,绘制标准曲线。

3. 水样中硫化物的检测

取实际水样 5.00mL 于 50mL 容量瓶中,用去离子水稀释至刻度,摇匀。准确移取 5.00mL 上述水样与 5.00mL 的氯化汞溶液（2μmol/L）混合,静置 3min,再加入 1.00mL 的石墨相氮化碳量子点,混匀。用荧光分光光度计,在与第 2 步相同条件下测试其荧光强度,平行测试三组。求出平均荧光强度值后带入标准曲线中,计算水样中硫化物含量。

五、思考题

1. 氯化汞溶液的浓度过大或过小对硫化物的检测会产生什么影响?
2. 在使用荧光分光光度计时,在操作上应注意什么?

实验二十一　土壤中有效磷的测定

一、实验目的

1. 了解光度法测定土壤中有效磷的原理及方法。
2. 熟悉分光光度计的使用方法。

二、实验原理

土壤中的磷大部分不能被植物直接吸收利用,易被吸收利用的有效磷通常含量很低。土壤中有效磷含量是指能被当季作物吸收的磷量。

土壤中有效磷的测定方法有生物法、同位素法、阴离子交换树脂法及化学法等,其中应用最普遍的是化学法。它是用浸提剂提取土壤中的一部分有效磷。浸提剂种类很多,应根据各种土壤的性质进行选择。酸性土壤中磷酸铁和磷酸铝形态的有效磷可用酸性氟化铵提取,形成氟铝化铵和氟铁化铵配合物,少量的钙离子形成氟化钙沉淀,磷酸根离子被提取到溶液中来。石灰性土壤则采用碳酸氢钠溶液浸取。

在含磷的溶液中加入钼酸铵,在一定酸度条件下,溶液中的磷酸与钼酸配位形成黄色的磷钼杂合酸——磷钼黄。

$$H_3PO_4 + 12H_2MoO_4 \rightleftharpoons H_3[PMo_{12}O_{40}] + 12H_2O$$

在适宜的试剂浓度下加入适当的还原剂（$SnCl_2$ 或抗坏血酸）,使磷钼酸中的一部分 Mo(Ⅶ) 还原为 Mo(Ⅵ),生成磷钼蓝（磷钼杂多蓝）——$H_3PO_4 \cdot 10MoO_3 \cdot Mo_2O_5$ 或 $H_3PO_4 \cdot 8MoO_3 \cdot 2Mo_2O_5$。在一定的浓度范围内,蓝色的深度与磷含量成正比,这是钼蓝比色法的基础。

三、主要仪器及试剂

仪器:分光光度计;25mL 容量瓶;50mL 锥形瓶;1mL、2mL 和 5mL 吸量管;振荡机；漏斗；无磷滤纸；分析天平。

试剂：0.5mol/L HCl 溶液；1mol/L NH_4F 溶液；100g/L H_3BO_3 溶液。

提取剂：分别移取 15mL 1mol/L NH_4F 溶液和 25mL 0.5mol/L HCl 溶液，加入 460mL 去离子水中，配制成 0.03mol/L NH_4F-0.025mol/L HCl 溶液。

15g/L 钼酸铵-3.5mol/L HCl 溶液：溶解 15g 钼酸铵于 300mL 去离子水中，加热至 60℃左右，如有沉淀，将溶液过滤，待溶液冷却后，慢慢加入 350mL 10mol/L HCl 溶液，并用玻璃棒迅速搅动，待溶液冷却至室温，用去离子水稀释至 1L，充分摇匀，储存于棕色瓶中，放置时间不得超过两个月。

25g/L 氯化亚锡溶液：称取氯化亚锡 2.5g 溶于 10mL 浓 HCl 中，溶解后加入 90mL 去离子水，混合均匀后置于棕色瓶中，此溶液现配现用。

50μg/mL 磷标准溶液：准确称取 105℃烘干的 KH_2PO_4 0.2195g，溶解于 400mL 水中，加浓 H_2SO_4 5mL（防止溶液长真菌），转入 1L 容量瓶中，加水稀释至刻度，摇匀，稀释 10 倍，此溶液不易久存。

四、实验步骤

1. 土壤样品预处理

称取风干土壤样品 1g（精确至 0.01g）于 50～100mL 小塑料瓶中，加入 0.03mol/L NH_4F-0.025mol/L HCl 溶液 20mL，稍摇匀立即放在振荡机上，振荡 30min。用无磷滤纸过滤，滤液承接于盛有 15 滴 100g/L H_3BO_3 溶液的 50mL 锥形瓶中（加 H_3BO_3 防止 F^- 对显色的干扰和腐蚀玻璃仪器），摇动瓶内溶液。

2. 土壤中有效磷的测定

准确移取上述土壤滤液 5～10mL（精确至 0.01mL）于 25mL 容量瓶中，用吸量管加入 15g/L 钼酸铵-盐酸溶液 5mL，加入去离子水定容，摇匀。

3. 工作曲线的制作

分别准确移取 5μg/mL 磷标准溶液 0mL、1.00mL、2.00mL、3.00mL、4.00mL、5.00mL 于 6 个 25mL 容量瓶中，加入 NH_4F-HCl 溶液 5～10mL（按所取滤液体积而定，精确至 0.01mL），用吸量管加钼酸铵-盐酸溶液 5mL，并滴加 25g/L 氯化亚锡 3 滴，摇动后，至溶液有深蓝色出现，用水稀释至刻度，摇匀，放置 15min，与土样溶液同时显色，测其吸光度。

以磷的浓度为横坐标，相应的吸光度为纵坐标，绘制标准曲线，并从标准曲线上查出土样中磷的含量。

五、注意事项

用氯化亚锡作还原剂生成磷钼盐，溶液的颜色不够稳定，必须严格控制比色时间，一般在显色后的 15～20min 内颜色较为稳定，显色后准确放置 15min 后立即比色，并在 5min 内完成比色操作。

六、思考题

1. 试述本实验测定磷的基本原理。
2. 测定吸光度为什么一般选择在最大吸收波长下进行？

实验二十二　食用油中不饱和脂肪酸的测定

一、实验目的

1. 了解食用油的组成和基本性质。
2. 掌握油脂碘值的测定方法，巩固氧化还原滴定基本操作。

二、实验原理

人体所需的必需脂肪酸主要来源于植物油尤其是食用油。构成食用油的脂肪酸从结构上可分为饱和脂肪酸和不饱和脂肪酸，不饱和脂肪酸又可以根据所含的双键数目不同分为单不饱和脂肪酸和多不饱和脂肪酸，单不饱和脂肪酸主要是油酸，多不饱和脂肪酸主要是亚油酸、亚麻酸、花生四烯酸、二十碳五烯酸（EPA）、二十二碳六烯酸（DHA）。大量实验研究发现，多不饱和脂肪酸是细胞膜的重要组成，对生命有机体的激素代谢和许多酶的活性起着调控作用，并能够降低心血管疾病发病率，抑制乳腺肿瘤和前列腺增生，延缓免疫功能衰退等。不饱和脂肪酸主要存在于橄榄油、芥花籽油、红花籽油、葵花籽油、玉米油、核桃油和大豆油等油中。

不饱和脂肪酸常用碘值来衡量，碘值是指 100g 物质中所能吸收（加成）的碘的质量（g）。碘值是油脂的重要特征之一，碘值越高，表明不饱和脂肪酸的含量越多，根据碘值的大小，可以鉴定油脂的不饱和程度，并以此衡量油脂的属性，例如：碘值大于 130 的油脂属于干性油脂；碘值小于 100 的油脂属于不干性油脂；碘值介于 100～130 之间的油脂属于半干性油脂。

本实验采用碘酊法测定碘值。碘乙醇溶液和水作用生成次碘酸，次碘酸和乙醇反应生成新生态碘，再和油脂中的不饱和脂肪酸起加成反应，剩余的碘，以硫代硫酸钠标准溶液滴定。其反应式如下：

$$I_2 + H_2O \rightleftharpoons HIO + HI$$
$$2HIO + C_2H_5OH \rightleftharpoons I_2 + CH_3CHO + 2H_2O$$
$$R_1CH=CHR_2 + 2HIO + C_2H_5OH \rightleftharpoons R_1CHI-CHIR_2 + CH_3CHO + 2H_2O$$
$$I_2 + 2Na_2S_2O_3 \rightleftharpoons 2NaI + Na_2S_4O_6$$

样品的碘值按下式计算

$$IV = \frac{0.1269 \times c_{Na_2S_2O_3}(V_0 - V_1)}{m} \times 100$$

式中　V_0——空白实验消耗的硫代硫酸钠标准溶液体积，mL；
　　　V_1——样品消耗的硫代硫酸钠标准溶液体积，mL；
　　　m——试样的质量，g。

0.1269 是 I_2 的毫摩尔质量，g/mmol。

三、仪器与药品

仪器：分析天平；恒温水浴；25mL 碘量瓶；25mL 移液管；25mL 酸式滴定管；100mL 量筒。

药品：无水乙醇；碘乙醇溶液（溶解分析纯碘片 25g 于 1L 95%乙醇中，放置 10d 后使用）；$Na_2S_2O_3$ 标准溶液（0.01mol/L）；淀粉指示剂。

样品：橄榄油、芝麻油、玉米油、大豆油。

四、实验步骤

称取 0.47~0.49 g 食用油于碘量瓶中，加 15mL 无水乙醇，使样品完全溶解，如果不易溶解，置于水浴中加温到 50~60℃ 至完全溶解，冷却。精确移取 0.2mol/L 碘乙醇溶液 20.00mL，注入已完全溶解并彻底冷却的样品液中，加水 100mL，塞紧瓶塞，充分摇荡，使成乳浊状，放置阴凉处 5min。然后以 0.1mol/L $Na_2S_2O_3$ 标准溶液滴定至浅黄色，加 1% 淀粉 1mL，继续滴定至蓝色消失为终点，消耗 $Na_2S_2O_3$ 标准溶液体积计为 V_1。不加食用油，同上法做空白试验，消耗 $Na_2S_2O_3$ 标准溶液体积计为 V_0。每个样品平行测定三次。

五、注意事项

1. 淀粉指示剂应在滴定到接近终点时加入，否则淀粉的螺旋腔与碘结合，造成误差。
2. 每个样品的 2 次测定值之差应不超过算术平均值的 0.5%。
3. 除本实验采用的化学滴定法外，还可采用气相色谱、液相色谱等方法检测食用油中不饱和脂肪酸的含量，如具备相关条件，可用多种方法进行检测并进行结果对比。

六、思考题

1. 不饱和脂肪酸碘值测定的方法还有哪些？请对比各方法之间的异同点。
2. 评价食用油质量还有哪些指标？

实验二十三　白酒中甲醇含量的测定

一、实验目的

1. 掌握白酒中甲醇含量测定的基本原理和方法。
2. 熟悉白酒中甲醇的卫生限量标准。

二、实验原理

甲醇是有毒化工产品，对人体有剧烈毒性。它对于视神经危害尤为严重，会引起视力模糊、眼疼、视力减退甚至失明。白酒中甲醇来自酿酒原辅料（薯干、马铃薯、水果、糠麸等）中的果胶，在蒸煮过程中果胶中的半乳糖醛酸甲酯分子中的甲氧基分解成甲醇。国家标准规定：凡是以各种谷类为原料制成的白酒，甲醇的含量不得超过 0.4g/L，以薯类为原料制成的白酒，则不得超过 1.2g/L。甲醇在磷酸介质中被高锰酸钾氧化为甲醛，甲醛与希夫（Schiff）试剂（亚硫酸钠品红溶液）反应后溶液呈蓝紫色，反应式如下。

1. 氧化：$5CH_3OH + 2KMnO_4 + 4H_3PO_4 = 5HCHO + 2KH_2PO_4 + 2MnHPO_4 + 8H_2O$
2. 去除有色物质：

$$5H_2C_2O_4 + 2KMnO_4 + 3H_2SO_4 = 2MnSO_4 + K_2SO_4 + 10CO_2\uparrow + 8H_2O$$

$$H_2C_2O_4 + MnO_2 + H_2SO_4 = MnSO_4 + 2CO_2\uparrow + 2H_2O$$

3. 显色反应：

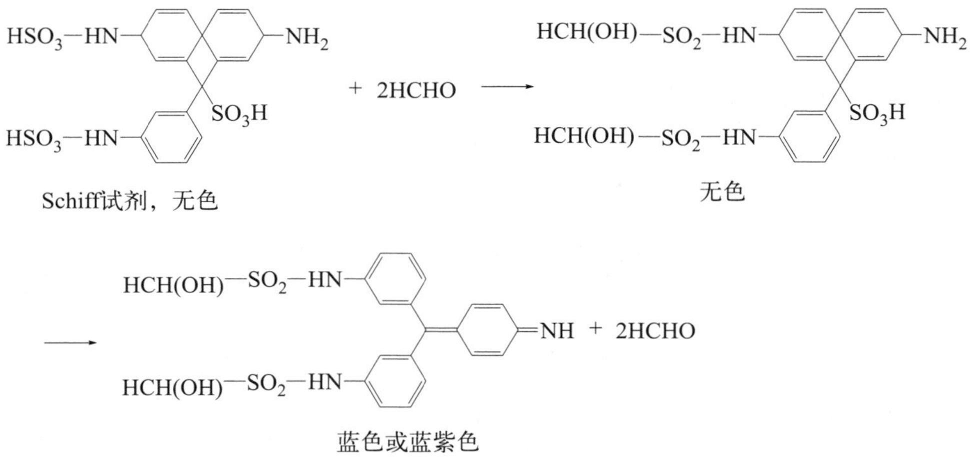

$$HSO_3-HN-C_6H_3-C(SO_3H)(C_6H_4)-C_6H_3-NH_2 + 2HCHO \longrightarrow HCH(OH)-SO_2-HN-C_6H_3-C(SO_3H)(C_6H_4)-C_6H_3-NH_2$$

Schiff试剂，无色 → 无色

$$\longrightarrow [HCH(OH)-SO_2-HN-C_6H_4]_2C=C_6H_4=NH + 2HCHO$$

蓝色或蓝紫色

在一定酸度下，甲醛所形成的蓝紫色不易褪色，而其他醛类形成的蓝紫色很容易消失，利用此反应测定甲醇含量。

三、仪器与试剂

仪器：分光光度计；恒温水浴箱；1mL、2mL、5mL、10mL 移液管；100mL 容量瓶；25ml 具塞比色管。

高锰酸钾-磷酸溶液：称取 3g 高锰酸钾，加入 15mL 85% 磷酸溶液及 70mL 水的混合液中，待高锰酸钾溶解后用水定容至 100mL。贮于棕色瓶中备用。

草酸-硫酸溶液：称取 5g 无水草酸（$H_2C_2O_4$）或 7g 含 2 个结晶水的草酸（$H_2C_2O_4 \cdot 2H_2O$），溶于 1:1 冷硫酸中，并用 1:1 冷硫酸定容至 100mL。混匀后，贮于棕色瓶中备用。

品红-亚硫酸溶液（Schiff 试剂）：称取 0.1g 研细的碱性品红，分次加水（80℃）共 60mL，边加水边研磨使其溶解，待其充分溶解后滤于 100mL 容量瓶中，冷却后加 10mL（10%）亚硫酸钠溶液，1mL 盐酸，再加水至刻度，充分混匀，放置过夜。如溶液有颜色，可加少量活性炭搅拌后过滤，贮于棕色瓶中，置暗处保存。溶液呈红色时应弃去重新配制。

甲醇标准溶液：准确称取 1.000g 甲醇（相当于 1.27mL）置于装有少量去离子水的 100mL 容量瓶中，加水稀释至刻度，混匀。此溶液每毫升含有甲醇 10mg，低温保存。

无甲醇无甲醛的乙醇：取 300mL 无水乙醇，加高锰酸钾少许，振摇后放置 24h，蒸馏，最初和最后的 1/10 蒸馏液弃去，收集中间蒸馏部分。

四、实验步骤

根据待测白酒中含乙醇多少适当取样（含乙醇 30% 取 1.0mL；40% 取 0.8mL；50% 取 0.6mL；60% 取 0.5mL）于 25mL 具塞比色管中。吸取 0.0、0.20、0.40、0.60、0.80、1.00mL 甲醇标准溶液分别置于 25mL 具塞比色管中。各加入 0.3mL 无甲醇无甲醛的乙醇。于样品管及标准管中各加水至 5mL，混匀，各管加入 2mL 高锰酸钾-磷酸溶液，混匀，放置

10min。各管加 2mL 草酸-硫酸溶液,混匀后静置,使溶液褪色。各管再加入 5mL 品红亚硫酸溶液,混匀,于20℃以上静置 0.5h。以 0 管为参比,于 590nm 波长处测吸光度,与标准曲线比较定量。

五、数据记录及结果处理

1. 标准曲线

用于绘制标准曲线的数据见表 4-2。

表 4-2　数据记录表

管号	0	1	2	3	4	5	样品 1	样品 2
甲醇标准液/mL	0.0	0.2	0.4	0.6	0.8	1.0	—	—
酒样/mL	—	—	—	—	—	—	0.6	0.6
甲醇含量/mg	0.0	2.0	4.0	6.0	8.0	10.0		
吸光度								

2. 甲醇含量

$$X = \frac{m}{V \times 1000} \times 100\%$$

式中　X——样品中甲醇的含量,g/(100mL);

　　　m——测定样品中所含的甲醇相当于标准甲醇的质量,mg;

　　　V——样品取样体积,mL。

六、注意事项

1. 亚硫酸品红溶液呈红色时应重新配制,新配制的亚硫酸品红溶液放冰箱中 24～48h 后再用为好。

2. 酒样和标准溶液中的乙醇浓度对吸光度有一定影响,故样品与标准管中乙醇含量要基本相等。

3. 除本实验采用的化学显色法外,还可采用气相色谱、液相色谱等方法检测白酒中的甲醇含量,如具备相关条件,可用多种方法进行检测并进行结果对比。

七、思考题

1. 我国食品卫生标准规定白酒中甲醇的允许含量是多少?过量甲醇对人体有何种危害?
2. 为什么加 Schiff 试剂前要除去过量的 $KMnO_4$?

附　录

一、定性分析试液的配制方法

（一）阳离子试液（含阳离子 10g/L）

阳离子	配制方法
Na^+	37g $NaNO_3$ 溶于水，稀释至1L
K^+	26g KNO_3 溶于水，稀释至1L
NH_4^+	44g NH_4NO_3 溶于水，稀释至1L
Mg^{2+}	106g $Mg(NO_3)_2 \cdot 6H_2O$ 溶于水，稀释至1L
Ca^{2+}	60g $Ca(NO_3)_2 \cdot 4H_2O$ 溶于水，稀释至1L
Sr^{2+}	32g $Sr(NO_3)_2 \cdot 4H_2O$ 溶于水，稀释至1L
Ba^{2+}	19g $Ba(NO_3)_2$ 溶于水，稀释至1L
Al^{3+}	139g $Al(NO_3)_3 \cdot 9H_2O$ 加 1:1 HNO_3 10mL，用水稀释至1L
Pb^{2+}	16g $Pb(NO_3)_2$ 加 1:1 HNO_3 10mL，用水稀释至1L
Cr^{3+}	77g $Cr(NO_3)_3 \cdot 9H_2O$ 溶于水，稀释至1L
Mn^{2+}	53g $Mn(NO_3)_2 \cdot 6H_2O$ 加 1:1 HNO_3 5mL，用水稀释至1L
Fe^{2+}	70g $(NH_4)_2SO_4 \cdot FeSO_4 \cdot 6H_2O$ 加 1:1 H_2SO_4 20mL，用水稀释至1L
Fe^{3+}	72g $Fe(NO_3)_3 \cdot 9H_2O$ 加 1:1 HNO_3 20mL，用水稀释至1L
Co^{2+}	50g $Co(NO_3)_2 \cdot 6H_2O$ 溶于水，稀释至1L
Ni^{2+}	50g $Ni(NO_3)_2 \cdot 6H_2O$ 溶于水，稀释至1L
Cu^{2+}	38g $Cu(NO_3)_2 \cdot 3H_2O$ 加 1:1 HNO_3 5mL，用水稀释至1L
Ag^+	16g $AgNO_3$ 溶于水，稀释至1L
Zn^{2+}	46g $Zn(NO_3)_2 \cdot 6H_2O$ 加 1:1 HNO_3 5mL，用水稀释至1L
Hg^{2+}	17g $Hg(NO_3)_2 \cdot H_2O$ 加 1:1 HNO_3 20mL，用水稀释至1L
Sn^{4+}	22g $SnCl_4$ 加 1:1 HCl 溶解，并用该酸稀释至1L

（二）阴离子试液（含阴离子 10g/L）

阴离子	配制方法
CO_3^{2-}	48g $Na_2CO_3 \cdot 10H_2O$ 溶于水，稀释至1L
NO_3^-	14g $NaNO_3$ 溶于水，稀释至1L
PO_4^{3-}	38g $Na_2HPO_4 \cdot 12H_2O$ 溶于水，稀释至1L
SO_4^{2-}	34g $Na_2SO_4 \cdot 10H_2O$ 溶于水，稀释至1L
SO_3^{2-}	16g Na_2SO_3 溶于水，稀释至1L
$S_2O_3^{2-}$	22g $Na_2S_2O_3 \cdot 5H_2O$ 溶于水，稀释至1L

续表

阴离子	配制方法
S^{2-}	75g $Na_2S \cdot 9H_2O$ 溶于水,稀释至 1L
Cl^-	17g NaCl 溶于水,稀释至 1L
I^-	13g KI 溶于水,稀释至 1L
CrO_4^{2-}	17g K_2CrO_4 溶于水,稀释至 1L

二、特殊试剂的配制方法

1. 酚酞（w 为 0.01）指示剂：溶解 1g 酚酞于 90mL 酒精与 10mL 水的混合液中。

2. 百里酚蓝和甲酚红混合指示剂：取 3 份 w 为 0.001 的百里酚蓝酒精溶液与 1 份 w 为 0.001 的甲酚红溶液混合均匀（在混合前一定要溶解完全）。

3. 淀粉（w 为 0.005）溶液：在盛有 5g 可溶性淀粉与 100mg 氯化锌的烧杯中，加入少量水，搅匀。把得到的糊状物倒入约 1L 正在沸腾的水中，搅匀并煮沸至完全透明。淀粉溶液最好现用现配。

4. 二苯胺磺酸钠（w 为 0.005）：称取 0.5g 二苯胺磺酸钠溶解于 100mL 水中，如溶液浑浊，可滴加少量 HCl 溶液。

5. 铬黑 T 指示剂：1g 铬黑 T 与 100g 无水 Na_2SO_4 固体混合，研磨均匀，放入干燥的磨口瓶中，保存于干燥器内。该指示剂也可配成 w 为 0.005 的溶液使用，配制方法如下：0.5g 铬黑 T 加 10mL 三乙醇胺和 90mL 乙醇，充分搅拌使其溶解完全。配制的溶液不宜久放。

6. 钙指示剂：钙指示剂与固体无水 Na_2SO_4 以 2∶100 比例混合，研磨均匀，放入干燥棕色瓶中，保存于干燥器内。或配成 w 为 0.005 的溶液使用（最好用新配制的）。配制方法与铬黑 T 类似。

7. 甲基红（w 为 0.001）：溶 0.1g 甲基红于 60mL 酒精中，加水稀释至 100mL。

8. 镁试剂Ⅰ：溶 0.001g 对硝基苯偶氮间苯二酚于 100mL 1mol/L NaOH 溶液中。

9. 铝试剂（w 为 0.002）：溶 0.2g 铝试剂于 100mL 水中。

10. 奈斯勒试剂：将 11.5g HgI_2 及 8 g KI 溶于水中稀释至 50mL，加入 6mol/L NaOH 50mL 静置后取清液贮于棕色瓶中。

11. 醋酸铀酰锌：溶解 10g $UO_2(Ac)_2 \cdot 2H_2O$ 于 6mL w 为 0.30 的 HAc 中，略微加热使其溶解，稀释至 50mL（溶液 A）。另溶解 30g $Zn(Ac)_2 \cdot 2H_2O$ 于 6mL w 为 0.30 的 HAc 中，搅动后稀释到 50mL（溶液 B）。将这两种溶液加热至 70℃后混合，静置 24h，取其澄清溶液贮于棕色瓶中。

12. 钼酸铵试剂（w 为 0.05）：5g$(NH_4)_2MoO_4$ 加 5mL 浓 HNO_3，加水至 100mL。

13. 磺基水杨酸（w 为 0.10）：10g 磺基水杨酸溶于 65mL 水中，加入 35mL 2mol/L NaOH，摇匀。

14. 铁铵矾 $(NH_4)Fe(SO_4)_2 \cdot 12H_2O$（$w$ 为 0.40）：铁铵矾的饱和水溶液加浓 HNO_3 至溶液变清。

15. 硫代乙酰胺（w 为 0.05）：溶解 5g 硫代乙酰胺于 100mL 水中，如浑浊须过滤。

16. 钴亚硝酸钠试剂：溶解 $NaNO_2$ 23g 于 50mL 水中，加 6mol/L HAc 16.5mL 及 $Co(NO_3)_2 \cdot 6H_2O$ 3g，静置过夜，过滤或取其清液，稀释至 100mL 贮存于棕色瓶中。每隔

四星期重新配置。或直接加六硝基合钴酸钠固体于水中,至溶液为深红色即可使用。

17. 邻菲咯啉指示剂（w 为 0.0025）：0.25g 邻菲咯啉加几滴 6mol/L H_2SO_4 溶于 100mL 水中。

18. 硫氰酸汞铵 $(NH_4)_2[Hg(SCN)_4]$：溶 8g $HgCl_2$ 和 9g NH_4SCN 于 100mL 水中。

19. 氯化亚锡（1mol/L）：溶 23g $SnCl_2 \cdot 2H_2O$ 于 34mL 浓 HCl 中,加水稀释至 100mL。

20. 甲基橙（w 为 0.001）：溶解 0.1g 甲基橙于 100mL 水中,必要时加以过滤。

21. 银氨溶液：溶解 1.7g $AgNO_3$ 于 17mL 浓氨水中,再用去离子水稀释至 1L。

22. 碘化钾-亚硫酸钠溶液：将 50g KI 和 200g $Na_2SO_3 \cdot 7H_2O$ 溶于 1000mL 水中。

23. α-萘胺：0.3g α-萘胺与 20mL 水煮沸,在所得溶液中加 150mL 2mol/L HAc。

24. 斐林试剂：①溶解 3.5g 分析纯的 $CuSO_4 \cdot 5H_2O$ 于含有数滴 H_2SO_4 的去离子水中,稀释溶至 50mL；②溶解 7g NaOH 及 17.5g 酒石酸钾钠于 40mL 水中,稀释溶液至 50mL；使用前把等体积的溶液②加入溶液①中,同时需充分搅拌。

25. 品红（w 为 0.001）溶液：将 0.1g 品红溶于 100mL 水中。

三、常用标准溶液的配制与标定

（一）直接配制的标准溶液

标准溶液	配制方法（均使用容量瓶）
0.05000mol/L Na_2CO_3	5.300g 基准 Na_2CO_3 溶于去离子水中（去二氧化碳）,稀释至 1L
0.05000mol/L $Na_2C_2O_4$	6.700g 基准 $Na_2C_2O_4$ 溶于去离子水中,稀释至 1L
0.01700mol/L $K_2Cr_2O_7$	5.001g 基准 $K_2Cr_2O_7$,用离子水溶解,稀释至 1L
0.1000mol/L NaCl	5.844g 基准 NaCl 溶于去离子水中,稀释至 1L
0.02500mol/L As_2O_3	4.946g 基准 As_2O_3、15g Na_2CO_3 在加热下溶于 150mL 去离子水,加 25mL 0.5mol/L H_2SO_4,稀释至 1L
0.01000mol/L $CaCl_2$	一级 $CaCO_3$ 在 110℃下干燥,称取 1.001g,用少量稀盐酸溶解,煮沸赶去二氧化碳后稀释至 1L
0.01000mol/L $ZnCl_2$	0.6538g 基准 Zn 加少量稀盐酸溶解,加几滴溴水,煮沸赶尽过剩的溴,稀释至 1L
0.1000mol/L 邻苯二甲酸氢钾	20.423g 基准邻苯二甲酸氢钾溶于去二氧化碳的去离子水中,稀释至 1L

（二）需要标定的标准溶液

标准溶液	配制方法	标定方法	
		实验步骤	指示剂
0.1mol/L HCl	浓 HCl 10mL 加水稀释至 1L	用本溶液滴定 25mL 0.05000mol/L Na_2CO_3。近终点时煮沸赶走 CO_2,冷却,滴定至终点	甲基橙
0.1mol/L NaOH	5g 分析纯 NaOH 溶于 5mL 去离子水中,离心沉降,用干燥的滴管取上层清液,用去二氧化碳的去离子水稀释至 1L	精确称取 2~2.5g 基准氨基磺酸,用容量瓶稀释至 250mL,取 25mL 用本溶液滴定	甲基橙
0.05mol/L $H_2C_2O_4$	6.4g $H_2C_2O_4 \cdot 2H_2O$ 加水稀释至 1L	用上面标定好的 NaOH 滴定	酚酞

续表

标准溶液	配制方法	标定方法 实验步骤	指示剂
0.02mol/L $KMnO_4$	约 3.3g $KMnO_4$ 溶于 1L 去离子水,煮沸 1～2 h,放置过夜,用四号玻璃砂漏斗过滤,贮于棕色瓶中,避光保存	取 25mL 0.05000mol/L $Na_2C_2O_4$,加 25mL 去离子水,10mL 9mol/L H_2SO_4,加热到 60～70℃,用本溶液滴定,近终点时逐滴加入至微红,30 s 不褪色即为终点	自身指示剂
0.1mol/L $FeSO_4$	28g $FeSO_4 \cdot 7H_2O$ 加水 300mL,浓 H_2SO_4 300mL,稀释至 1L	用本溶液 25mL,加 25mL 0.5mol/L H_2SO_4,5mL 85% H_3PO_4,用上面 0.02mol/L $KMnO_4$ 进行滴定	$KMnO_4$
0.1mol/L $(NH_4)_2Fe(SO_4)_2$	40g $(NH_4)_2Fe(SO_4)_2 \cdot 6H_2O$ 溶于 300mL 2mol/L H_2SO_4 中并稀释至 1L	标定方法同 0.1mol/L $FeSO_4$ 溶液的标定方法	$KMnO_4$
0.05mol/L I_2	12.7g I_2 和 40g KI,溶于去离子水并稀释至 1L	a. 本溶液 25mL,用本表中 0.1mol/L $Na_2S_2O_3$ 滴定,指示剂:淀粉 b. 取 25mL 0.02500mol/L As_2O_3 稀释一倍,加 1g $NaHCO_3$,用本溶液滴定	淀粉
0.1mol/L $Na_2S_2O_3$	25g $Na_2S_2O_3 \cdot 5H_2O$ 用 1L 煮沸冷却后的去离子水溶解,加少量 Na_2CO_3 贮于棕色瓶中,放置 1～2 天后标定	取 25mL 0.01700mol/L $K_2Cr_2O_7$ 加 5mL 3mol/L H_2SO_4 和 2g KI,以本溶液滴定	淀粉(要进行空白实验)
0.1mol/L $AgNO_3$	17g $AgNO_3$ 加水溶解并稀释至 1L,贮于棕色瓶中,避光保存	取 25mL 0.1000mol/L NaCl,加 25mL 水,5mL 2% 的糊精,用本溶液滴定	荧光黄
0.1mol/L KSCN	9.7g KSCN 溶于煮沸并冷却的去离子水中,稀释至 1L	取本表中 0.1mol/L $AgNO_3$ 25mL,加 5mL 6mol/L HNO_3,用本溶液滴定	$(NH_4)Fe(SO_4)_2 \cdot 12H_2O$ 饱和溶液 1mL
0.1mol/L NH_4SCN	8g NH_4SCN 加水溶解并稀释至 1L	同上	$(NH_4)Fe(SO_4)_2 \cdot 12H_2O$ 饱和溶液 1mL
0.01mol/L EDTA	3.8g EDTA·2Na·$2H_2O$ 溶于水并稀释至 1L	取 25mL 0.01000mol/L $CaCl_2$ 或 0.01000mol/L $ZnCl_2$ 溶液,加 0.1mol/L NaOH 中和后,加 3mL pH=10 的缓冲溶液(70g NH_4Cl 和 570mL $NH_3 \cdot H_2O$ 稀释至 1L)和 1mL 0.1mol/L Mg-EDTA,用本溶液滴定	铬黑T

四、常见离子鉴定方法

(一) 常见阳离子的鉴定方法

阳离子	实验步骤及注意事项
Na^+	取 3 滴 Na^+ 试液,加 12 滴醋酸铀酰锌试剂,放置数分钟,用玻璃棒摩擦器壁,淡黄色的晶状沉淀出现,指示有 Na^+ $3UO_2^{2+} + Zn^{2+} + Na^+ + 9Ac^- + 9H_2O \Longrightarrow 3UO_2(Ac)_2 \cdot Zn(Ac)_2 \cdot NaAc \cdot 9H_2O \downarrow$ 注意:1. 鉴定宜在中性或 HAc 酸性溶液中进行,强酸、强碱均能使试剂分解; 2. 大量 K^+ 存在时,会干扰鉴定;Ag^+、Hg^{2+}、Sb^{3+} 有干扰;PO_4^{3-}、AsO_4^{3-} 能使试剂分解

续表

阳离子	实验步骤及注意事项
K^+	加入5滴六硝基合钴酸钠($Na_3[Co(NO_2)_6]$)溶液于4滴K^+试液中,放置片刻,若有黄色的沉淀$K_2Na[Co(NO_2)_6]$析出,说明K^+存在 注意:1. 鉴定宜在中性、微酸性溶液中进行,因强酸、强碱均能使$[Co(NO_2)_6]^{3-}$分解; 2. NH_4^+与试剂生成橙色沉淀而干扰,但在沸水浴中加热1~2min后,$(NH_4)_2Na[Co(NO_2)_6]$完全分解,而$K_2Na[Co(NO_2)_6]$不变
NH_4^+	气室法:用干燥、洁净的表面皿两块(一大一小),在大的一块表面皿中心放5滴NH_4^+试液,再加5滴6mol/L NaOH溶液,混合均匀。在小的一块表面皿中心粘附一小条润湿的酚酞试纸,盖在大的表面皿上形成气室。将此气室放在水浴上微热2min,酚酞试纸变红,证明有NH_4^+,这是NH_4^+的特征反应
Ca^{2+}	取2滴Ca^{2+}试液,滴加饱和$(NH_4)_2C_2O_4$溶液,有白色的CaC_2O_4沉淀形成,证明有Ca^{2+} 注意:1. 反应宜在HAc酸性、中性、碱性溶液中进行; 2. Mg^{2+}、Sr^{2+}、Ba^{2+}的存在干扰反应,但MgC_2O_4溶于醋酸,Sr^{2+}、Ba^{2+}应在鉴定前除去
Mg^{2+}	取4滴Mg^{2+}试液,加入4滴2mol/L NaOH溶液,2滴镁试剂Ⅰ,沉淀呈天蓝色,证明有Mg^{2+} 注意:1. 反应宜在碱性溶液中进行,NH_4^+浓度过大会影响鉴定,故在鉴定前加碱煮沸,除去NH_4^+ 2. Ag^+、Hg^{2+}、Hg_2^{2+}、Cu^{2+}、Co^{2+}、Ni^{2+}、Mn^{2+}、Cr^{3+}、Fe^{3+}及大量Ca^{2+}干扰反应,应预先分离
Ba^{2+}	取2滴Ba^{2+}试液,加1滴0.1mol/L K_2CrO_4溶液,有黄色沉淀生成,证明有Ba^{2+} 鉴定宜在HAc-NH_4Ac的缓冲溶液中进行
Al^{3+}	2滴Al^{3+}试液,4~6滴水和4滴3mol/L NH_4Ac及4滴铝试剂搅拌,微热,加6mol/L $NH_3 \cdot H_2O$至碱性,红色沉淀不消失,证明有Al^{3+} 注意:1. 鉴定宜在HAc-NH_4Ac的缓冲溶液中进行; 2. Cr^{3+}、Fe^{3+}、Bi^{3+}、Cu^{2+}、Ca^{2+}对鉴定有干扰,但加氨水后,Cr^{3+}、Cu^{2+}生成的红色化合物分解,$(NH_4)_2CO_3$加入可使Ca^{2+}生成$CaCO_3$、Fe^{3+}、Bi^{3+}、Cu^{2+}可预先和NaOH形成沉淀而分离
Sn^{2+}	取4滴Sn^{2+}试液,加2滴0.1mol/L $HgCl_2$溶液,生成白色沉淀,证明有Sn^{2+} 注意:若白色沉淀生成后,颜色迅速变灰、变黑,这是由于Hg_2Cl_2进一步被还原为Hg
Pb^{2+}	取2滴Pb^{2+}试液,加2滴0.1mol/L K_2CrO_4溶液,生成黄色沉淀,表示有Pb^{2+} 注意:1. 鉴定在HAc溶液中进行,因为沉淀在强酸强碱中均可溶解; 2. Ba^{2+}、Bi^{3+}、Hg^{2+}、Ag^+等干扰鉴定
Cr^{3+}	向3滴Cr^{3+}试液中滴加6mol/L NaOH溶液直至生成的沉淀溶解,搅动后加4滴w为0.03的H_2O_2,水浴加热,待溶液变为黄色后,继续加热; Cr^{3+}的氧化需在强碱性条件下进行,而形成$PbCrO_4$的反应,须在弱酸性(HAc)溶液中进行
Mn^{2+}	取1滴Mn^{2+}试液,加10滴水,5滴2mol/L HNO_3溶液,然后加少许$NaBiO_3$(s),搅拌,水浴加热,形成紫色溶液,证明有Mn^{2+} 注意:1. 鉴定反应可在HNO_3或者H_2SO_4酸性溶液中进行; 2. 还原剂(Cl^-、Br^-、I^-、H_2O_2等)干扰反应
Fe^{3+}	1. 取2滴Fe^{3+}试液,放在白滴板上,加2滴2mol/L HCl及2滴$K_4[Fe(CN)_6]$溶液,生成蓝色沉淀,证明有Fe^{3+} 注意:1. 反应在酸性溶液中进行; 2. 大量存在Cu^{2+}、Co^{2+}、Ni^{2+}等离子,有干扰,需分离后再作鉴定 2. 取2滴Fe^{3+}试液,加2滴0.5mol/L NH_4SCN溶液,形成血红色溶液,表示有Fe^{3+} 注意:1. F^-、H_3PO_4、$H_2C_2O_4$、酒石酸、柠檬酸等能与Fe^{3+}形成稳定的配合物而干扰反应; 2. Co^{2+}、Ni^{2+}、Cr^{3+}和铜盐,因离子有色,会降低检验Fe^{3+}的灵敏度

续表

阳离子	实验步骤及注意事项
Fe^{2+}	1. 取 2 滴 Fe^{2+} 试液在白色滴板上,加 2 滴 2mol/L HCl 及 1 滴 $K_3[Fe(CN)_6]$ 溶液,出现蓝色沉淀,表示有 Fe^{2+} 注意:反应在酸性溶液中进行 2. 取 1 滴 Fe^{2+} 试液,加几滴 w 为 0.0025 的邻菲咯啉溶液,生成橘红色溶液,证明有 Fe^{2+} 注意:反应在微酸性溶液中进行,选择性和灵敏度均较好
Co^{2+}	取 2~3 滴 Co^{2+} 试剂,加饱和 NH_4SCN 溶液 12 滴,加 8~9 滴戊醇溶液,振荡,静置,有机层呈蓝绿色,证明有 Co^{2+} 注意:1. 反应需用浓 NH_4SCN 溶液; 2. Fe^{3+} 有干扰,加 NaF 掩蔽,大量 Cu^{2+} 也有干扰
Ni^{2+}	取 2 滴 Ni^{2+} 试液,2 滴 6mol/L 氨水,2 滴二乙酰二肟溶液放在白色点滴板上,凹槽四周形成红色沉淀证明有 Ni^{2+} 注意:1. 反应在氨性溶液中进行,合适的酸度 pH=5~10; 2. Fe^{2+}、Fe^{3+}、Cu^{2+}、Co^{2+}、Cr^{3+}、Mn^{2+} 有干扰,可加柠檬酸或酒石酸掩蔽
Cu^{2+}	2 滴 Cu^{2+} 试液,2 滴 6mol/L HAc 酸化,2 滴 $K_4[Fe(CN)_6]$ 溶液混合,红棕色沉淀出现则证明有 Cu^{2+} 注意:1. 反应宜在中性或弱酸性溶液中进行; 2. Fe^{3+} 及大量的 Co^{2+}、Ni^{2+} 会干扰
Ag^+	2 滴 Ag^+ 试液和 2 滴 2mol/L HCl 混匀,水浴加热,离心分离,在沉淀上加 4 滴 6mol/L 氨水,沉淀溶解,再加 6mol/L HNO_3 酸化,白色沉淀重又出现,证明有 Ag^+
Zn^{2+}	取 1 滴 Zn^{2+} 试液,用 2mol/L HAc 酸化,加入等体积的 $(NH_4)_2Hg(SCN)_4$ 溶液,生成白色沉淀则有 Zn^{2+} 注意:1. 反应在中性或微酸性溶液中进行; 2. 少量 Co^{2+}、Cu^{2+} 存在,形成蓝紫色混晶,有利于观察,但含量大时有干扰。Fe^{3+} 有干扰
Hg^{2+}	取 1 滴 Hg^{2+} 试液,加 1mol/L KI 溶液,使生成的沉淀完全溶解后,加 2 滴 $KI-Na_2SO_3$ 溶液,2~3 滴 Cu^{2+} 溶液,生成橘黄色沉淀则有 Hg^{2+} CuI 是还原剂,须考虑到氧化剂(Ag^+、Fe^{3+} 等)的干扰

(二)常见阴离子的鉴定方法

阴离子	实验步骤及注意事项
Cl^-	取 1 滴 Cl^- 试液,加 6mol/L HNO_3 酸化,滴加 0.1mol/L $AgNO_3$ 至沉淀完全,离心分离,在沉淀上加 3~4 滴银氨溶液,混匀加热至沉淀溶解,再加 6mol/L HNO_3 酸化,有白色沉淀生成,说明有 Cl^- 存在
Br^-	取 2 滴 Br^- 试液,加入数滴四氯化碳溶液后滴加氯水,振荡,有机层呈橙红或橙黄色,说明 Br^- 存在 氯水宜边滴加边振荡,若氯水过量会生成 BrCl,使有机层呈淡黄色
I^-	取 2 滴 I^- 试液,加入数滴四氯化碳溶液后滴加氯水,振荡,有机层呈紫色,说明 I^- 存在 注意:1. 反应宜在酸性、中性或弱碱性条件下进行; 2. 过量氯水将 I_2 氧化成 IO_3^-,有机层紫色将褪去
SO_4^{2-}	取 3 滴 SO_4^{2-} 试液,滴加 6mol/L HCl 酸化,加 3 滴 0.1mol/L $BaCl_2$ 溶液,生成白色沉淀,说明 SO_4^{2-} 的存在
SO_3^{2-}	向 2 滴饱和硫酸锌溶液中加 2 滴 0.1mol/L $K_4[Fe(CN)_6]$ 溶液,生成白色沉淀,继续加 2 滴 $Na_2[Fe(CN)_5NO]$,2 滴中性 SO_3^{2-} 试液,白色沉淀转换为红色沉淀$(Zn_2[Fe(CN)_5NO]SO_3)$,说明 SO_3^{2-} 的存在 注意:1. 酸能使沉淀消失,酸性溶液需用氨水中和; 2. 硫离子 S^{2-} 有干扰,需要预先除去

续表

阴离子	实验步骤及注意事项
$S_2O_3^{2-}$	1. 取 3 滴 $S_2O_3^{2-}$ 试液,加 3 滴 2mol/L HCl 溶液,微热出现白色浑浊,说明 $S_2O_3^{2-}$ 的存在; 2. 取 3 滴 $S_2O_3^{2-}$ 试液,加 8 滴 0.1mol/L $AgNO_3$ 溶液,振荡,若生成的白色沉淀迅速变黄→棕→黑色,说明 $S_2O_3^{2-}$ 存在。 注意:1. S^{2-} 存在时,$AgNO_3$ 溶液加入时,由于有 Ag_2S 生成,干扰观察 $Ag_2S_2O_3$ 沉淀的颜色变化; 2. $Ag_2S_2O_3$ 可溶于过量可溶性硫代硫酸盐溶液中
S^{2-}	1. 取 5 滴 S^{2-} 试液,稀硫酸酸化后用 $Pb(Ac)_2$ 试纸检验生成的气体,试纸变黑即说明存在 S^{2-}; 2. 取 3 滴 S^{2-} 试液于白滴板上,加 3 滴 $Na_2[Fe(CN)_5NO]$,溶液变紫色说明 S^{2-} 的存在(反应在碱性条件下进行)
CO_3^{2-}	1. 浓度较大的 CO_3^{2-} 溶液用 1mol/L HCl 酸化,产生的二氧化碳气体使澄清石灰水或氢氧化钡溶液变浑浊,说明 CO_3^{2-} 存在 2. 当 CO_3^{2-} 含量较少或同时存在其他能与酸反应生成气体的物质时,用氢氧化钡气瓶法检验: 取出滴管,在玻璃瓶中加少量 CO_3^{2-} 试样,从滴管上口加一滴饱和氢氧化钡溶液,再加入 6 滴 1mol/L HCl,立即将滴管插入瓶中并塞紧,轻敲瓶底静置数分钟,溶液浑浊则证明有 CO_3^{2-} 注意:1. 如果氢氧化钡溶液浑浊度不大,可能是吸收空气中的二氧化碳所致,需进行空白实验比较; 2. 如果试液中含有 $S_2O_3^{2-}$ 和 SO_3^{2-},为消除其干扰,预先加入数滴 H_2O_2 将它们氧化为 SO_4^{2-},再检验 CO_3^{2-}
NO_3^-	1. 当 NO_2^- 同时存在时,4 滴试液中加 8 滴 12mol/L H_2SO_4 和 4 滴 α-萘胺,生成淡紫红色化合物即证明了 NO_3^- 的存在 2. 当 NO_2^- 不存在时,用稍过量的 6mol/L HAc 酸化 4 滴 NO_3^- 试液,加少许镁片搅动,NO_3^- 还原为 NO_2^-;取 4 滴上层清液,按照下面 NO_2^- 的鉴别方法鉴定
NO_2^-	向 HAc 酸化的 4 滴试液中加 1mol/L KI 和 CCl_4,振荡,有机层呈紫红色,证明 NO_2^- 的存在
PO_4^{3-}	取 2 滴 PO_4^{3-} 试液,加 8~10 滴钼酸铵试剂,用玻璃棒摩擦内壁,生成黄色钼酸铵沉淀,说明 PO_4^{3-} 存在 $$PO_4^{3-}+3NH_4^++12MoO_4^{2-}+24H^+ = (NH_4)_3P(Mo_3O_{10})_4+12H_2O$$ 注意:1. 沉淀可以溶于碱或氨水中,所以反应要在酸性条件下进行; 2. 如果存在还原剂,可使 Mo^{VI} 还原为钼蓝而使溶液呈深蓝色,需要预先除去; 3. 与 PO_3^-、$P_2O_7^{4-}$ 的冷溶液不反应,煮沸后由于生成 PO_4^{3-} 而进一步生成黄色沉淀

参考文献

[1] 商少明,汪云,刘瑛. 无机及分析化学实验. 3版. 北京:化学工业出版社,2019.
[2] 陈立钢,廖丽霞,牛娜. 分析化学实验. 北京:科学出版社,2015.
[3] 李琳,陈爱霞. 无机化学实验. 北京:化学工业出版社,2020.
[4] 杨芳,郑文杰. 无机化学实验(中英双语版). 北京:化学工业出版社,2020.
[5] 叶艳青,庞鹏飞,汪正良. 无机与分析化学. 北京:化学工业出版社,2021.
[6] 张丽影,王茹. 双语无机及分析化学实验. 沈阳:辽宁大学出版社,2014.
[7] 苏文,杨辉. 柠檬提取液生物合成纳米硒及其抗氧化性[J]. 精细化工,2020,37(11):2266-2272.
[8] 陈少东,陈福北,杨帮乐,等. 几种食用油中不饱和脂肪酸和皂化值的测定研究[J]. 化工技术与开发,2011,40(10):53-55.
[9] 续炎,马浩迪,徐孝楠,等. 细菌胞外液生物合成纳米硒的理化性质及抗菌活性研究[J]. 化学研究与应用,2023,35(12):2828-2835.